# Sun Cheese

## Lessons From Regenerative Farming That Can Help You Rewrite Your Future

**Siobhan M. Griffin**

Published by Restoracene

***Sun Cheese:***
***Lessons From Regenerative Farming***
***That Can Help You Rewrite Your Future***

*First Edition*

**Siobhan M. Griffin**

**Published by Restoracene, August 2026.**

**www.restoracene.com**



ISBN: 979-8-9961010-4-7 (paperback)

ISBN: 979-8-9961010-6-1 (hardcover)

ISBN: 979-8-9961010-5-4 (ebook)

Library of Congress Control Number: 2026917997



Copyedited by: Marie Griffin, Trudi Cameron, Anne Downes, Eric Hubler, Erik Hane, and Rob Coulson.

Cover and Interior Design by: Scott Moden

For

Alana, Bruce,

and all

Grandchildren

on Planet Earth.

**Jasmine Aburto, a "grandchild" to me and a nature lover who helped me discover so much insect life at Raindance Farm.** Her mom, Rafaela, was herdswoman at Raindance Farm for 11 years.

Photo by Siobhan Griffin, July 15, 2015

# Contents

# List of Figures and Tables

# Foreword

Few things make people pause more than a simple sentence: "A thousand years ago the Earth produced more nutrition than it does today." This shocks people. They scratch their chins and don't know how to process it.

Truth irritates. My dad and mom purchased the most worn-out, gullied rockpile of a farm property in Virginia's Shenandoah Valley in 1961. The rich alfisoils that greeted European settlers in the 1750s eroded to subsoil and then shale rock over the next 200 years. The pastures barely fed 10 cows. Today, those same pastures feed 100 cows and we've never purchased a bag of chemical fertilizer or planted a seed. Ponds, butterflies, rich soil, and earthworms adorn this previously desecrated spot of creation.

How is that possible? Early on, my dad discovered André Voisin, the French father of what he called "rational grazing," which holistic management guru Allan Savory today credits as the epiphany in his own journey to restorative land management and decision making.

A Smithsonian study recently found all nine known types of bumblebees indigenous to Virginia happily buzzing on our farm. I can't remember ever seeing a wild turkey on our farm as a child. Now we have flocks of 30 or more. A deer was a rarity; now they're almost a nuisance. My experience rattles both the conservatives and the environmentalists. The conservatives don't see how compost could heal what chemicals destroyed. The environmentalists don't see how commercial farming can increase biodiversity and soil depth. Few things are as entertaining as questioning what most folks know to be true.

Here are a few:

1. Ecological destruction started with Monsanto and the chemical companies.
2. Herbivores are environmental nightmares.
3. People starve because we don't produce enough food.
4. Carbon credits to plant pine trees will help the environment.
5. Individually we're powerless to change the Earth's trajectory.

*Sun Cheese: Lessons From Regenerative Farming That Can Help You Rewrite Your Future* takes on these and many more conventional "knowns," casting a wide net over the many threads of human planetary exploitation. This is not a warm, fuzzy book. In fact, I bristled at a couple of things that seemed like partisan political swipes. But one of our problems these days is looking for things we can disagree on rather than things we can agree on.

For example, we could disagree with many of the dating assumptions in this book. But the fact remains that Libya provided baboons and lions to the Roman Colosseum. These animals don't live in deserts; they live in jungles. Do you see jungles in Libya today? The Greek historian Homer walked across northern Africa and said he never left the shade of a tree. Today, that is the expanding Sahara.

Siobhan masterfully explains, anthropologically and archeologically, that virtually all the barren deserts of the world were once heavily vegetated and populated with megafauna like mammoths and saber-toothed tigers. The record exists in fossils, cave drawings, and sediment excavations. It's an uncomfortable read because it questions all the assumptions and bogeymen of the modern conventional environmental message. And it dares to poke at conventional libertarianism and amoral capitalism. She's like me, having the ability to make everyone unhappy.

My favorite theme is that she doesn't let any of us off the hook of responsibility for creating a Restoracene epoch. In our "blame those guys" temptation, we love personal escapism. But Siobhan puts the blame on us and puts the onus for redemption on us. As in, you and me. Not that other guy. Looking in the mirror.

Anyone who reads this book will find things that seem wrong, sensationalized, or not germane to the ecological issue. Like income inequality. But stick with her argument and you'll see the dots connected. What may seem like a rabbit trail is part of a whole.

And that's the point of the book. You can't just plant a tree. You have to plant the right tree in the right place for the right reason, and maybe you should plant grass and install an electric fence instead of the tree. Most books like this conclude with a recipe. After explaining and describing the dead end we're on, you're hoping for a nice, clean solution. But in keeping with holistic management, Siobhan refuses to insult us with a recipe.

She reverts to each of us doing what a holistic exercise would lead us to do. I deeply appreciate her call to stop our luxury exploitation just because a paycheck says we can buy something or travel somewhere. Like she found on her farm, a piece of ground provides enough soul satisfaction, entertainment, discovery, and contentment. We don't need fashion, bright shiny objects, and more stuff. Minimalism can be highly motivational and enjoyable.

This book gifts us with a mandate to think. In a TikTok world disconnected from our ecological umbilical, the mandate to think is a revolutionary imperative: to think in ways we haven't thought about things we need to think about. In general, our ignorance is astounding. Siobhan spent three years deeply researching the scientific literature for this book. Her experiences on her own farm, interactions with her own property, and participation in various earth actions accentuate the urgency of her message.

The bottom line is this: Even if you throw out the parts of the book with which you disagree, you're left with a pile of evidence that humans have not treated our home kindly. And we continue to disrespect it in profound ways. What's more, I do. Even as supposedly caring and thoughtful as I am, I blanched with shame a couple of times while reading these words. Yes, some hypocrisy. Guilty. What about you? Be courageous enough to expose yourself to challenging thoughts. It'll do your soul good, and maybe add one more person to the Restoracene.

**Joel Salatin**

**Ground nesting bird nests appeared in great abundance on Raindance Farm pastures after switching to holistic planned grazing.**

Photo by Siobhan Griffin, July 15, 2015

# Introduction

When I visited Lahaina, Hawaii, in 2016, I saw wildfires waiting to happen all over the dry side of Maui. Farmers had stopped growing pineapple and other crops, and the land was growing copious amounts of tall, dry grass and brush. This made lovely tinder for a wildfire in such a dry climate. Over the past 40,000 years, our ancestors erased many species of megafauna that once were powerful natural forces keeping wildfire at bay. Now, communities all over the globe were seeing wildfire and floods rear their ugly heads. But it doesn't have to be this way. With better understanding of how nature managed wildfire, we can prevent many and reduce their severity.

The Lahaina Fire Incident Analysis Report issued by the Fire Safety Research Institute nicely supported my thesis that land management was to blame for many such fires. It identified unmanaged vegetation on state and fallow private land upwind of Lahaina as a major predisposing factor for the wildfire. More importantly, the report concluded that the threat could be addressed in cheap, practical ways such as grazing off the offending brush and grass.[1]

Then I watched the news as if it was on repeat when the LA fires inflicted their devastation in January 2025. The tragedy came as no surprise to me because I live in Southern California now during the summer. When I travel around our town near San Diego, every day I see wildfires waiting to happen on the hillsides as brush is allowed to build up there, too. Ironically, the same management of neglect sets up California residents for flooding also, as the area of bare soil increases underneath the bushes—a surefire sign of desertification. Still, publishers were rejecting my writing that predicted these fires because they said books about climate change do not sell. But this book is not about climate change. It is about people change.

1 Alkonis, Steve and Derek Kerber., "Lahaina Fire Incident Analysis Report," Fire Safety Research Institute (FSRI), September 14, 2024, 59–62. https://doi.org/10.60752/102376.26858962.

There is a dire need for people to change how we manage our lives and businesses to address the dual crises of biodiversity loss and climate disruption. Changing how people behave is a lot harder to achieve than restoring wildlife and calming climate volatility will be once people finally do realize the need to change for the sake of all future generations.

## Human Nature

Understanding the natural world is not enough to get us out of the environmental mess we are in. We must understand human nature too. We collectively have known about climate disruption for several decades, but fossil fuel consumption and pollution continue to increase. As climate-related disasters become a normal occurrence on the news and the planet reaches an extra two degrees of warming, it is clear we are not acting adequately to improve outcomes. Some people are even conveniently in denial that humans have anything to do with climate change. This should surprise no one. In my experience, people are unlikely to change how they behave until they change how they see themselves. The classic example is when an alcoholic realizes they are one. Only then do they see a reason to change how they act. To inspire responsible action toward the preservation of the biosphere that can influence climate, I reveal evidence that humans have been impacting the global climate for 64,000 years. This agency to change climate by how we manage nature is a double-edged sword, but it holds the key to a kinder future.

Many poor decisions to mitigate climate volatility are being implemented out of ignorance on such a widespread scale that it is putting everyone's future in jeopardy. For instance, monoculture forest plantations are being planted worldwide to sequester carbon, even though the forestry sector in Canada is a net emitter of greenhouse gases. There, wildfire ignites much of the carbon held in trees and releases it rapidly into the air. Astonishingly, if the reality of fire is included in the tally, the forestry sector in Canada emitted more greenhouse gases in 2023 than the country's fossil fuel industry, including dirty coal tar extraction! There is nothing natural about these wildfires, as you will learn in Chapter 13, and they will only get worse if we continue business as usual. Don't get me wrong—I love native forests and am planting one on my property. The world needs more trees. But they must be managed in nature's image to function for the benefit of humanity.

We can look to nature for guidance about how to address the root causes of many of humanity's problems, from wildfire to climate change. For instance, restoring the function of grasslands, especially in semiarid places like California, could store more carbon long-term than monoculture forestry, while also reducing wildfire risk.[2] Restoring the abundance of fish worldwide could speed up carbon sequestration in the ocean.[3]

So how do we know what actions are the best ones to take to keep life viable and this planet gorgeous? I don't tell you what to do, but you will know how to figure it out in your unique place and time after you read this book. My hopeful message is that you can have more autonomy over your own future than you ever imagined. But it requires intentional dreaming—and taking action to make your dream happen.

## Reconnecting With Nature

Rewriting your future requires imagining your landscape, including the land where your food is grown, as it needs to look to provide all the good things in life. Nature has always provided everything required for life, including more security than we have now from disasters like floods and wildfires. But today's generations have lost the connection to nature that once gave our ancestors intuitive understanding of how nature protects us, solves problems, and creates abundance. In the early chapters of *Sun Cheese* I take readers living in the glare of modern society, beyond their screens and cities to meet the forgotten ecosystems that grow their food and make their economies possible—and always have. We cannot restore ecosystems to perform well if we don't know what they're supposed to look like and how they should function. Pristine paleolithic landscapes offer highly productive templates that can inform the human creativity required to reimagine agroecosystems that are fit for purpose today and going forward for as long as we wish our species to endure.

Everywhere our human ancestors arrived for the first time, they met splendor akin to the forests and oceans in the Avatar series. These must be restored because fully functioning ecosystems feed our species and all others. The unique and giant paleolithic animals of Australia, such as *Diprotodon optatum*, were just a few examples of the wonders that the first people to arrive encountered. These great marsupials the size of small cars, and many other now-extinct megafauna, were essential ecosystem players that cocreated the green cloak of vegetation that kept climate wetter on the Australian continent. Grass and trees induced rain that

2 Pawlok Dass et al., "Grasslands May Be More Reliable Carbon Sinks than Forests in California," Environmental Research Letters 13, no. 7 (2018): Results and Disc., https://doi.org/10.1088/1748-9326/aacb39.

3 Gaël Mariani et al., "Let More Big Fish Sink: Fisheries Prevent Blue Carbon Sequestration—Half in Unprofitable Areas," Science Advances 6, no. 44 (2020): 1

filled megalakes teeming with life, even in the now-arid heart of Australia. After my readers explore the great breadth and inner workings of the biosphere with me, they will be able to see much deeper into the soul of their own local landscape. With hearts and minds open, their land can speak to them with clues about how to cultivate productive ecosystems that will feed, water, and shelter thousands of generations to come.

## The Holy Grail: Ecosystem Productivity

Ten thousand years ago, our blue/green planet was twice as productive as today. This has huge ramifications because in the absence of so much life, Earth has less capacity to cool itself now. I reveal some of the patterns and templates nature uses to keep her biosphere humming along as productively as possible. The intricate dance between all living things, including the megafaunal giants we have forgotten, wasn't merely spectacular to behold. Life produced the oxygen we breathe; built deep, fertile soils where we grow our food; and filled the vast underground aquifers we drink and irrigate from today. The production of these gifts from nature is coming to a grinding halt as humans rip apart the fabric of life and strip nature of her power.

Why are people collectively killing the golden goose that has provided everything required for life? How do we resurrect a natural capital inheritance for our children that instead is teeming with beautiful wildlife and is productive for humans? A prerequisite is to own up to our human role in the diminishment of nature. *Homo sapiens* have been effective killers for tens of thousands of years and you will learn how in Chapter 5, "The Rise of the Super Predator." Once we take responsibility for our total human impact on our planet, it becomes obvious that only we can fix the damage to our biosphere. After all, we are the ones devouring it. But it doesn't have to be this way. We can become caretakers of the living tapestry that cradles us from before birth to after death. We can be good ancestors—heroes to our grandchildren instead of generations to be cursed.

In the past, human success had allowed our ancestors to steal countless wonders from our generation as they sent over 150 species of amazing megafauna to oblivion. Today's generations are stealing our own children's inheritance at an accelerating rate. The abundance of wild animals and insects had plummeted 70 percent since I was a teenager! But we can be a positive force to restore landscapes so that they once again create rain and infiltrate water to replenish depleted aquifers that future generations will depend on.

We can rebuild lost topsoil faster than previously believed possible—and sequester meaningful quantities of carbon as we do it. I know we can because my family and I did these things on our farm, as I share in Chapters 10 and 11.

As much as I believe in the idea of growing food in ways that enhance wildlife where we live and farm, life in the ocean emerged in my research as a powerful force for cooling our planet that is equally worth protecting. People have been diminishing this power, including the "whale pump" that sequesters carbon, for well over 150 years. You will learn in Chapter 12, "Ocean Armageddon," that it doesn't have to be this way. Reviving these powerful forces is up to you and starts with holistic consumer decisions. What you buy to put on your dinner plate and even what you recycle to keep pollution out of waterways means life and death for ocean wildlife that cumulatively cool planet Earth.

So why the title *Sun Cheese*? Making farmstead cheese was the first holistic decision we considered on our organic farm. The explosion of biodiversity on our land that resulted from switching to holistic management inspired my research into pristine ecosystems and the writing of this book. Holistic management is not just for farmers. Managing all our lives and businesses holistically will allow people everywhere to make a living while looking after the ecosystems that provide everything required for life. This is a story of empowerment that I hope will inspire you to write your own future script so your family can enjoy security, peace, and abundance on cooler, cleaner, greener landscapes filled with life. My faith in humankind makes me believe that once you realize the need to manage your life holistically (simultaneously valuing nature and people while you make a living), you will create a future that is better for all our grandchildren and theirs.

"Know Thyself... "

— One of three inscriptions on the Temple of Apollo in the ancient Greek precinct of Delphi.

# Chapter 1

# A Tragic Story of Hope

Gindi sat spellbound on the limb of a tree. Underneath her, a vast herd of huge, snorting, bugling, sweating herbivorous marsupials, each the size of a rhino, was passing through the grove and slowly spilling out onto the grassy plains beyond. She gave her open-mouthed cousin, Myaru, a little nudge, smiled, and pointed at the pouch filled with tubers they had earlier dug out of the loose, friable soil. There were so many giant marsupials passing underneath their tree that she could not even see the end of the migrating herd. Gindi settled in for a long wait, thinking of the giant thunderbird (*Dromornis stirtoni*) egg they had eaten for breakfast, which would keep them satisfied until dinner.

The grove they sat in was teeming with life. In a nearby tree, large, colorful birds played bass in a symphony of birdsong. They, too, were patiently waiting for the herd to pass so they could swoop down to dine on trampled small marsupials and lizards, or maybe even an old, toothless *Diprotodon optatum* that had been left to die, unable to keep up with the herd. Above her, Gindi could see the sky darken with flocks of birds that followed the giant marsupials. Below, she noticed a mob of emus approaching, all looking to feast on the seemingly infinite insects being flushed out of their hiding places in the tall, lush grass by the tramping of heavy limbs. The wildlife concert was regularly punctuated by the loud quacking of flocks of giant ducks that ran noisily by, feasting on the plethora of aquatic plants in the nearby ponds frequented by giant platypuses and crocodiles. Gindi's people, too, added their voices to the wildlife symphony because they were one with all this life in a circular economy.

As the yellow sun slipped lower in the sky, the last of the herd finally were passing through the savanna. Gindi spotted her father and uncles peering out from behind trees on the opposite hillside. The men slunk out onto the plain to confront six *Diprotodon* females, some with babies, at the flank of the herd. The giant marsupials didn't run away; they quickly formed a defensive circle around their young just as they did when faced with groups of marsupial lions.

The young joeys scrambled into the pouches between their mothers' legs, pouches that were spacious enough to carry a human adult. Few predators could penetrate this defense without being trampled to death by the enraged mothers. But the men didn't need to get that close. They calmly moved apart to surround the group, and each man fitted his spear to his woomera—a spear-throwing device that magnified the power of his throw. Gindi's eldest uncle chose a well-fattened, barren young female and let his weapon fly from a safe distance. The others threw their spears in unison and dispatched the target game quickly and efficiently. Then, to avoid the wrath of the other angry females, they jogged over to the grove to greet the smiling girls with whoops of triumph. There would be a feast tonight! When the other *Diprotodons* saw the human predators retreating, they dispersed with their offspring, leaving one motionless animal behind.

Myaru's father told the girls to run to the riverside to ask the rest of the community to come and start butchering the carcass. He knew the girls would be safe from ferocious marsupial lions because he and his brothers had trapped and killed the last one known to be in the area. Creating safety and providing game meat for their families was men's work. Humans would be the dominant predator now for tens of thousands of years here, and in countless locations around the planet where humankind were making their mark. The men stayed behind to guard their prize from scavengers and to start a fire big enough to cook a feast of meat, herbs, and murnong (yam daisy) roots for all 82 people in their tribe.

The day before, Gindi's tribe had enjoyed a day of fishing and had feasted on a roasted fish that weighed as much as she did. Their landscape in what is now central Australia, 50,000 years ago, was one of abundance. Unlike the arid climate there today, soft rain often fell upon the girls' faces. Life there exhaled gray mist that clung to the trees that broke up vast expanses of grassland and marshes. On certain mornings, this mist would condense into a dome of cloud above multitudes of emus gathered on the grasslands. As in all regions of the world, our species' success and creative technology would change the face of this land slowly over time.

Migrating *Diprotodon*, mobs of emus (*Dromaius novaehollandiae*), giant short-faced kangaroos (*Procoptodon goliah*), and a plethora of other giant marsupials and reptiles kept busy fertilizing a green landscape. The plant material that these animals trampled along with their dung kept the soil soft, productive, and loaded with water-absorbing carbon to store seasonal rainwater. In response, the land rewarded the fertilizing herbivores with succulent grasses, flowers, shrubs, and trees to graze upon. Gindi didn't realize that this noisy green cloak of plants and animals covering the land induced the rain that filled the many chains of ponds, mighty rivers, and even the megalakes that were thousands of square kilometers in size, dotting her continent. Nor could she have ever imagined that the water seeping underground from her well-watered landscape was supplying the Great Artesian Basin, one of the largest and deepest aquifers globally, today extending to over 1.7 million $km^2$. How could she ever have guessed that 50,000 years later, this water would be pumped to the surface by people for drinking and irrigation in a parched land?

Gindi's distant relatives who lived at the time on the savannas of Eurasia, Arabia, and the Sahara of northern Africa witnessed similar scenes and symphonies. Only the characters changed. In all these places, bountiful, well-watered savanna landscapes were created and maintained by giant gardeners. These green lands had hippopotamuses and multiple species of elephant, including the straight-tusked *Palaeoloxodon antiquus*. In the northern reaches of these kingdoms roamed several species of mammoth (Mammuthus) and large rhinoceros, including woolly rhinos (*Coelodonta antiquitatis*). Specialized predators like saber-toothed cats in the genus *Homotherium* and large cave lions (*Panthera spelaea*) prowled across Europe. As herbivorous giants foraged for food, cared for their young, and interacted with their social groups as symbionts in the web of life, they fertilized soil and created food and habitat for other life. The result was rich, green, rain-inducing landscapes that trickled their precious, life-giving liquid over thousands of years into giant transboundary aquifers that people depend upon today, like the Lake Chad Basin (2.43 million $km^2$, covering parts of Chad, Niger, Central African Republic, Nigeria, Algeria, Sudan, Cameroon, and Libya) and the Arabian Aquifer System (1.86 million $km^2$, spanning Saudi Arabia, Iraq, Kuwait, Bahrain, Qatar, the United Arab Emirates, Oman, and Yemen).

This fossil water stored underground is now worth trillions of dollars, enabling agriculture and human settlement in what have since become dry regions. However, people are rapidly drawing down this fossil water to quench their crops and cities. Thirty percent of the world's aquifers are shrinking at an accelerating rate, especially in arid and semiarid regions.[4] As water continues to be unsustainably exploited, it will become priceless—but maybe it always was. A huge problem today is that the intact green ecosystems that facilitate water infiltration to recharge groundwater resources are not valued by modern generations.

4 Scott Jasechko et al., "Rapid Groundwater Decline and Some Cases of Recovery in Aquifers Globally," *Nature* 625, no. 7996 (2024): 715, https://doi.org/10.1038/s41586-023-06879-8.

They continue to be depleted and no longer replenish the aquifers as fast as people use the water. Now, with so much rain-inducing and water-absorbent savanna, forest, and swampland all but wiped off the face of the Earth, can contemporary *Homo sapiens* learn how to replenish the aquifers of our planet once again for the sake of our grandchildren? Is better management of our planet's biosphere the key to a livable future?

## Natural Capital

Water, fertile topsoil, fossil fuels, and forests are rapidly being used up in all corners of the planet and exchanged for money that is disproportionately enriching a tiny percentage of people. As soil and ecosystem degradation continues, we are seeing more frequent and more severe flooding and wildfire events. The area suitable for growing food is shrinking fast. Restoring ecosystem integrity and productivity is essential so our descendants can benefit from the life-giving gifts that nature provides. These can range from the coral and shell beaches created by marine life, to the climate-moderating effect of long-term carbon sequestration that only occurs in oceans that are filled with life.

Since the root cause of the linked crises of biodiversity loss and climate change is how we manage our lives, economies, and nature, that must be where we start to rewrite the future we really want for ourselves and our children. We can never go back to the paradise our early human ancestors inherited, but nature has tremendous power to heal if given a chance. Now is the time to create the conditions for this to happen and to give evolution a chance again to change the face of the planet. After all, what is the point of having children if we do not pass on to them a planet worth inhabiting? We can steward the capacity of land and seascapes to provide them with enough clean water, quality soil for growing food, and plenty of green plants to produce the very oxygen they will breathe.

## The Revelation

Agriculture is the most significant contributor to declining ecosystems today because it has evolved into systems that focus on high production from one crop species, using poisons to exterminate all other species in the process. When chemical companies were coming up with products to sell to farmers as "solutions" to problems like insects eating crops, they rarely considered the knock-on effects on people and the environment, or even the economic impact on people outside their company. Multiply this kind of reductionist decision-making over thousands of products, and the result is a dangerous and destructive food system.

In 2016, I was camping at Sunset State Beach, California, in a hired RV with a longboard on the roof. I was on a low-budget farmer's vacation, and it was fun … until 5 a.m. one dark morning, when I was rudely awakened by the stench of farm chemicals. Even low levels of chemical exposure make me feel very ill, so I packed up the van immediately and hightailed it out of the campground.

As I pulled onto a local road in the early morning darkness, I was stunned to see large farm equipment with powerful lights illuminating the field across the road. A fog of poisonous soil fumigant was pouring out from under it. Fumigants are liquid pesticides that form a gas upon contact with the soil. Even more shocking were the backlit silhouettes of farm workers pulling on a roll of plastic about 10 feet wide attached to the back of the vehicle as a chemical fog swirled around their heads. They were laying down white plastic to cover even more of the already plastic-clad valley to try to keep the poisonous gas in the soil, where it was meant to kill a teeming microcosm of soil life down to one meter deep.

The use of fumigants allows farmers to grow continuous monocrops like strawberries without needing to rotate crops to prevent pathogens from building, as was common practice in the past. For an organic farmer like me, this scene was like one from a nightmare. Destroying the extensive soil biology that I had carefully encouraged on my farm for three decades was contrary to my core principles. I wondered, how does widespread use of these poisons affect human health?

Fumigants are toxic to a broad range of organisms. According to a paper in the *Encyclopedia of Agriculture and Food Systems* in 2014, "To be effective, chemicals used for soil fumigation must be extremely toxic to a wide range of organisms encompassing actinomycetes, bacteria, fungi, oomycetes, insects, nematodes, and plants (both seeds and vegetative material)."[5] According to the United States Environmental Protection Agency (EPA), soil fumigation can pose serious health risks to farm workers and neighbors. They state on their website,

"As gases … fumigants move from the soil to the air at the application site and may move off site at concentrations that produce adverse health effects in people from hours to days after application. These health effects range from mild and reversible eye irritation to more severe and irreversible effects, depending on the fumigant and the level of exposure." [6]

---

5 D. O. Chellemi, "Plant Health Management: Soil Fumigation," *Encyclopedia of Agriculture and Food Systems* (2014): 456. https://doi.org/10.1016/B978-0-444-52512-3.00250-3.

6 United States Environmental Protection Agency, "What Are Soil Fumigants?," accessed January 15, 2024, https://www.epa. gov/soil-fumigants/what-are-soil-fumigants.

Californian farmers used 176 million lb. of pesticide and herbicide active ingredients on 98 million acres in 2023, according to the California Department of Pesticide Regulation (CDPR). In their most recent pesticide use report, the CDPR discloses that 32,471,609 lb. of active ingredients that are known carcinogens were applied as part of chemical spray mixes just in California,[7] supposedly one of the most regulated states in the US.

I am glad I got out of the campground quickly, but what about the families who live, work, and go to school nearby? How can fruits and vegetables grown with carcinogens be healthy for my family? I never would have cajoled my children into eating more fruits and vegetables if I had known they were grown using so many poisonous chemicals! How were we able to grow vegetables, fruits, nuts, eggs, meat, and dairy on our organic farm in New York State without sterilizing the soil or using synthetic pesticides? Why did we have thriving birds, lightning bugs, and butterflies on the same land that produced three quarters of a million pounds of nutrient-dense milk each year? It turns out the buying decisions of modern consumers play the biggest role in how food is produced, so everyone must be better informed for the sake of human and ecological health.

Organic farms around the world use crop rotations and other intensive management interventions to prevent pest buildup in soil. The same monoculture crops grown year on year are very vulnerable. As the ecologist Frans Vera aptly put it, "Agriculture and forestry create a limited monster from the total of biodiversity."[8] Sadly, many people are told that vegetables and fruits produced with cocktails of chemicals and planted under non-recycled plastic in poisoned, lifeless soils are healthier for them and the planet than the chemical-free food we grew on our family dairy farm, which also supported a wealth of wildlife.

## Biodiversity Loss

While many people are worrying about the dangers of climate change, a bigger but linked crisis is still largely overlooked. Humans are on the verge of causing the sixth mass extinction of wildlife, which has decreased in abundance by 73 percent since 1970, according to a 2022 report from the World Wildlife Fund.[9]

---

7 California Department of Pesticide Regulation, "Pesticide Use Reporting: 2023 Summary Data," Carcinogens: pounds applied, 4, https://www.cdpr.ca.gov/wp-content/uploads/2025/08/2023_carcinogens_pounds_applied.pdf.

8 F. W. M. Vera, *Grazing Ecology and Forest History* (CABI, 2000), 381.

9 WWF (World Wildlife Fund), *Living Planet Report* 2024, 32.

As superpredators, humans have become a force of nature unlike any other species on Earth. Over tens of thousands of years, Homo sapiens have displaced nonhumans from their ecological niches and taken over their resources. For example, the mighty bison herds of America that once numbered 60 million or more were systematically slaughtered by European settlers in the 1800s to a paltry several hundred individuals. Bison rangelands, where they had helped build deep topsoil for thousands of years, were appropriated by humans for farming and ranching.

In the 1970s, most wildlife population sizes were larger compared to today. Little did we know then that the abundance of wild animals my generation used to see on the TV show *Wild Kingdom*, and the fish and coral we saw Jacques Cousteau swim amongst on our TV screens in the 1970s, were mere remnants of the abundance that flourished tens of thousands of years ago alongside our Stone Age ancestors.[10] This is a big problem because animals fertilize primary producers like plants and algae and keep them healthy so they can produce organic matter for consumers further up the food chain to eat, a process called primary production. If plants, algae, and photosynthetic bacteria could not harvest solar energy and use it to turn $CO_2$ and water into edible, carbohydrate-dense biomass and oxygen, we and almost all the life around us wouldn't be here. The rate of biomass production, or "net primary productivity" (NPP), can be measured by the weight of biomass grown per unit of the Earth's surface area in a given amount of time. For instance, on my farm, our land produced about seven tonnes of grass per hectare per year.

Lush ecosystems create the natural capital our economies depend on, everything from fertile topsoil to forests, so it is worth pointing out some other basic concepts of ecology that run throughout this book. Consumers are creatures who consume plant and algal biomass and cycle nutrients that, in turn, fertilize plants. Larger animals are much more powerful at both eating and fertilizing plants wherever they roam in nature's original ecosystem templates. So, they serve to speed up how fast plants grow and harvest solar energy, a process known in ecology as energy turnover.

Energy turnover can be thought of as the "metabolic rate" of a whole ecosystem. For instance, if you spend the day on the couch watching a screen, your metabolic rate is low, and you don't need much food to maintain your weight because you are using little energy. But you also don't get much work done.

---

**10** Yinon M. Bar-On et al., "The Biomass Distribution on Earth," *Proceedings of the National Academy of Sciences of the United States of America* 115, no. 25 (2018): 6508, https://doi.org/10.1073/pnas.1711842115.

On the other hand, if you spend all morning mowing the lawn and trimming your hedges, you will require a big lunch to replace the energy you used because your metabolic rate was higher as you burned more calories. But you also got a lot of work accomplished. Fueled by the sun, cooperating life forms in pristine ecosystems accomplish a lot of work that benefits people, like growing food for us and our fellow earthling species. When energy turnover is higher, more food is fed to consumers and carbon is sequestered in soils faster. So, if ecosystem productivity increases, people and nonhuman life benefit more. In summary, ecosystem primary production is the size of the plant and algal "solar panels", and ecosystem turnover measures how efficiently these solar panels work.

The present-day biomass of large wild land mammals is only 8 percent of what it was between 100,000 and 742,000 years ago, before our species spread around the world. Numbers of the largest land mammals that ever lived—megafauna like elephants, mammoths, and woolly rhinos—have decreased from 1.1 billion to 80 million today, and the ecological impacts have been far-reaching. Energy turnover stimulated by the grazing and fertilizing effects of these large mammals has decreased accordingly by 92 percent. This has worsened ecosystem degradation.[11] The land degradation caused by gutting energy turnover in ecosystems widely across the planet ultimately would bleed the greenhouse gas $CO_2$ into the atmosphere and cause climate warming. Thus, primary productivity is the holy grail of ecosystem restoration and has profound implications for a livable climate and our future.

## Planetary Overshoot

All the Earth's ecosystems combined make up the biosphere, composed of all living things. To put the damage to the Earth's biosphere into perspective, the combined total carbon content embedded in various forms of life on Earth can be used as a measuring stick of relative abundance, because almost half the dry weight of animals and plants is carbon. Combined, plants contain 80 percent of the biosphere's carbon[12] (**Figure 1.1**) but their mass is only half of what it was before the start of human civilization.

**11** Juraj Bergman et al., "Worldwide Late Pleistocene and Early Holocene Population Declines in Extant Megafauna Are Asso-ciated with Homo Sapiens Expansion Rather than Climate Change," *Nature Communications* 14, no. 1 (2023): Consequenc-es of megafaunal decline, para. 3, https://doi.org/10.1038/s41467-023-43426-5.

**12** Bar-On et al., "Biomass Distribution," 6507, Fig 1.

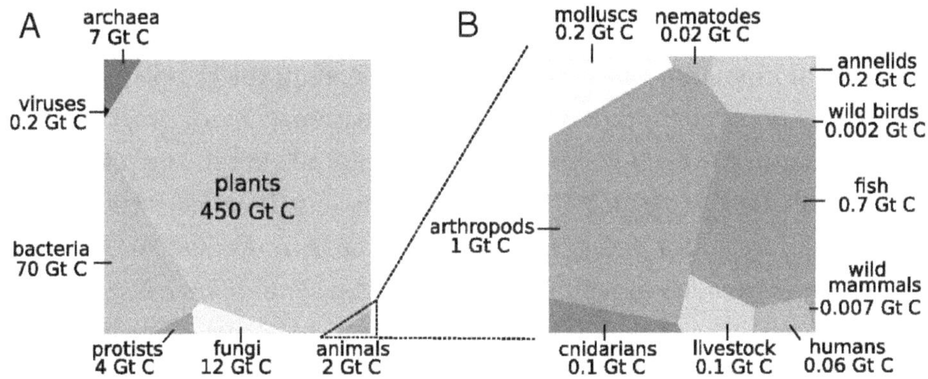


**Figure 1.1 Graphical representation of the global biomass distribution by taxa.**

Source: Bar-On et al., 2018.

Over the past two centuries, reductionist decisions made to rapidly exploit and consume vast amounts of carbon-dense fossil fuels for profit have allowed humans to escape the limitations of natural energy availability through photosynthesis. *Homo sapiens* took about 250,000 years to reach a global population of 1 billion in 1820, and just over 200 years to go from 1 billion to 8 billion.[13] As a result, the total weight of all mammals on Earth has increased approximately fourfold in the last 200 years, but only because of the biomass of people and their associated livestock.[14] Powered by ancient fossil energy, the population of humanity has exceeded the Earth's capacity to absorb our waste and pollution. Now, the rate we are pushing other nonhuman earthlings to extinction has exponentially accelerated even though no one wants to see other species go extinct. How can this juggernaut of species destruction be reversed?

Ignoring the contribution of other species to human economic systems can harm our own survival. Every species, from the largest elephant to the smallest soil microbe, plays a role in keeping the land and sea productive so we all can breathe, have fresh water, and eat. Sadly, now the biomass of dogs or sheep alone outweighs all wild mammals combined, including elephants.[15]

13 Joseph J. Merz, Pheobe Barnard, William E. Rees et al., "World Scientists' Warning: The Behavioural Crisis Driving Ecological Overshoot," *Science Progress* 106, no. 3 (2023): Human behavior drives overshoot, para. 2 https://doi. org/10.1177/00368504231201372.

14 Anthony D. Barnosky, "Megafauna Biomass Tradeoff as a Driver of Quaternary and Future Extinctions," *Proceedings of the National Academy of Sciences of the United States of America* 105 (supplement 1) (2008): 11545, fig. 5.https://doi. org/10.1073/pnas.0801918105.

15 Lior Greenspoon et al., "The Global Biomass of Wild Mammals," *Proceedings of the National Academy of Sciences of the United States of America* 120, no. 10 (2023): Disc., para. 6. https://doi.org/10.1073/pnas.2204892120.

## The Root Cause of Planetary Decline

At this pivotal moment in human history, we need all the people we've got to jumpstart a new, more productive era. A holistic management practitioner and educator in South Africa, Roland Kroon, has been training young people to use livestock to rewild the plains of that country. He has flown over endless miles of scrubland in western Zambia where, he told me in an August 2023 interview, many people are needed to restore these damaged semiarid ecosystems before they degrade further to the point that desertification becomes irreversible. He says, "You can drive for three or four hours and not see a single person, but the landscape is totally degraded because of lack of animals." He thinks that "people must realize the mess we are in and what caused that mess: it is decision-making. And it is reductionist thinking."

Our ancestors did not want to lose iconic megafaunal species that accompanied them on the plains and in the forests where they lived. Surely, they wanted their grandchildren to benefit from the migration of hooved, finned, and winged game that arrived seasonally at their doorstep to provide food. Why wouldn't they want their grandchildren to have the opportunity to harvest the eggs of the giant thunderbird? They probably revered these animals for the sustenance they provided.

No one I know wants to experience the floods that occurred after Hurricane Katrina, or the catastrophic wildfire that plagued Los Angeles in 2024 and 2025. No one wants to think too hard about the permanent loss of iconic species like the northern white rhinoceros (*Ceratotherium simum cottoni*) during their watch. Nor do we want to see a huge exodus of people fleeing their spent, desertified homelands. None of us asked for this heartless future that is unfolding, but all people have inadvertently created it by making reductionist decisions every day. The problem was, and still is, the reductionist way that *Homo sapiens* manage nature, societies, and economies by focusing on just one problem, want, or need at a time. However, as a species, we can develop new decision-making frameworks that are purpose-built, and then we can start a restorative era.

## The "Restoracene"

*Homo sapiens* can evolve to a more advanced way of making all our important decisions, from "What is for dinner?" to "Should I take a joy ride to the edge of space?" By taking this leap of consciousness, we can begin a new era in our human story, which I have termed the "Restoracene."

My personal journey down the restorative path started with the first holistic decision we considered on our farm, making a farmstead cheese we named Sun Cheese. We had to live it to believe it, as green plants, wildlife, and birds gradually became more abundant on Raindance Farm, our family's dairy farm. We started to learn how to help nature build up natural capital again in the form of topsoil and groundwater. Our holistic land management practices built deeper, more carbon-dense soil that absorbed more rainwater and allowed it to trickle down through the soil profile to eventually emerge as clear spring water instead of rushing off in a muddy torrent.

Many young people today are losing hope that they can make a difference as climate change and species extinction appear to be unstoppable. But there is no reason people everywhere cannot change how we all manage our lives, economies, and the natural resources around us for better outcomes. But for change in behavior to happen, people must first own the fact that we are destroying our life support system as we hurtle through deep, dark space. Only then will we see the need to change. The ecosystem restoration that began to take place on our farm made me realize that everyone can make a difference for future generations and pay our good fortune of abundance forward. This gives me profound hope. In his 1916 tome, *A Book Lover's Holidays in the Open*, Theodore Roosevelt wrote,

> If in a given community unchecked popular rule means unlimited waste and destruction of the natural resources—soil, fertility, water-power, forests, game, wildlife generally—which by right belong as much to subsequent generations as to the present generation, then it is sure proof that the present generation is not yet really fit for self-control; that it is not yet really fit to exercise the high and responsible privilege of a rule which shall be both by the people and for the people.[16]

**16** Theodore Roosevelt, *A Book Lover's Holidays in the Open* (Charles Scribner's Sons, 1916), 319.

## Chapter 2

# Practical Solutions

Migrants from afar gather to work in the sweltering greenhouses in southern Spain during the European winters. They grow vegetables for European tables. Here, near San Silvestre, Almeria, Spain, white plastic film covers the entire landscape on a scale visible from space. You can see it for yourself on Google Maps. Greenhouse plastic is an impermeable film that disrupts the water cycle, making clean, fresh water unavailable for the last remnants of a savanna ecosystem that once teemed with animals and plants and provided food for people on its own.

Imagine how much better the migrant workers' lives and your life would be if they could make a living restoring ecosystems and water cycles and greening the dusty, often conflict-riddled homelands they hail from. What if Europeans decided to buy more nutrient-dense seasonal vegetables from local producers who honor productive green spaces? What if people in semiarid regions like California, southern Spain, and the Sahel could reverse desertification, prevent wildfires, and contribute to climate cooling? If enough people restored the productive green agroecosystems where they live, it would reduce the economically damaging extreme weather and sea level rise globally. But we cannot solve the climate crisis without giving everyone the know-how to heal their ecosystems locally. How is this done?

## Restoring the Biosphere to Cool Planet Earth

The Restoracene begins when humans decide to pull Spaceship Earth up from the ecological nosedive she is enduring by switching to ecological economies that value and look after the natural capital that present and future generations need. If restoring biology to mitigate climate change sounds like a ridiculous oversimplification, it couldn't be farther from the truth. The complexity of nature is almost infinite and constantly evolving. When you throw in numerous human factors such as culture and the need for people to make a living, complexity overwhelms human decision-making. Holistic management frameworks help people, businesses, and policymakers navigate this complexity and will propel us into the Restoracene.

Pyramids and stone ruins stand in testament to the many ancient civilizations that once thrived on productive land that has since turned to desert. In the following chapters, I will reveal the role humans have played in the desertification of huge swaths of the planet, including the Sahara and the interior of Australia. Despite not intending to, people have weakened their life-supporting ecosystems for eons. Negligently, we continue to manage nature in the same reductionist manner as our cave-dwelling ancestors did, causing massive topsoil loss across the globe and species extinctions.

In the American Corn Belt, for instance, tillage has caused topsoil to wash or blow away at unsustainable levels. There is little evidence that conservation practices since the Dust Bowl have slowed this rate of soil loss.[17] Massive dust storms have returned. A recent study found that soil erosion on US cropland in 2022 was 4.72 tonnes per hectare per year and is projected to get worse as climate change progresses. Soil erosion was highest where corn, soybean, sorghum, and wheat were grown and lowest on forestlands, shrublands, and rangelands. Corn and soybeans are cultivated on 17 percent of US land, but they account for 44 percent of the country's soil erosion.[18] If the way people make decisions was fit for purpose, we wouldn't be sitting on an ecological knife-edge, seeing soil and species disappear, and the global climate becoming untethered from the stability our species enjoyed during our entire evolution. Getting back on the right track will take effort, and we must be equipped with solid ecological information.

Other effective holistic decision-making frameworks may develop, but we already have a shovel-ready framework developed by ecologist Allan Savory in collaboration with many experts and proven to work over decades of practice. Educators affiliated with the Savory Institute and Holistic Management International teach both rich and poor people the basics of ecology and the decision-making process that can empower everyone to contribute to a better future. Farmers and business owners are adopting frameworks to restore green, more profitable, climate-cooling agroecosystems and to be more climate-friendly. Some holistic ranchers are even reversing desertification. But a holistic mindset is required by everyone if the damaging human impact on our planet is to be reversed.

17 Evan A. Thaler et al., "Rates of Historical Anthropogenic Soil Erosion in the Midwestern United States," *Earth's Future* 10, no. 3 (2022):10, https://doi.org/10.1029/2021EF002396.

18 Shahab Aldin Shojaeezadeh et al., "Soil Erosion in the United States: Present and Future (2020–2050)," *CATENA* 242 (2024): sec.3.1, para.2,108074, https://doi.org/10.1016/j.catena.2024.108074.

Help from consumers, especially in the US, is critical because farmers make up less than 2 percent of the population. Everyone who eats food, drinks water, and breathes air has a stake in this movement to create a better future and can make a difference, starting with their consumer decisions. But even urban people need a deeper understanding of the ecology of restoration if they are to make informed decisions.

My baby-boomer generation has failed badly to pass on a better planet to the next. Humanity's only hope is that younger generations do better. Allan Savory's daughter, Sarah, has recently completed a children's book on healthier, empowering ways to make decisions. *Never Lose Sight of Your Elephant and Your Decisions Will Change the World!* is available from the Savory Institute and aims to change the trajectory of readers' futures, young and old, and help jump-start the Restoracene.

## Changing Our Food Production Paradigms

One reason for worldwide damage to the most productive soils is the land-sparing approach that justifies poisoning or stripping all life from farmland except the desired crop, in the hope that high productivity will allow other areas to be set aside and remain pristine. In practice, however, not much wild area is being spared because of the sheer number of people to feed. Certainly not enough to keep the planet cool.

In response to the Earth Summit held in Rio de Janeiro, Brazil, in June 1992, protected areas doubled globally as policymakers recognized the strong correlation between human impacts and biodiversity loss. But protected areas currently only cover 14.7 percent of the world's land area, and these areas are increasingly under pressure from rapidly increasing human populations. Only 42 percent of this "protected" land area is not intensely impacted by human activity, including intensive farming, grazing, major road infrastructure, and even towns and villages.[19] Human impact on protected areas is greatest in Europe, but every country needs to prioritize habitat for the planet's two billion other species. Protecting the last few remnants of the lungs of our planet, tropical rainforests, and the productivity of the oceans, where phytoplankton produce about half of the oxygen we breathe, really is a matter of life or death.

19 Kendall R. Jones et al., "One-Third of Global Protected Land Is under Intense Human Pressure," *Science* 360, no. 6390 (2018): 788, https://doi.org/10.1126/science.aap9565.

Many people like the idea of setting aside land to rewild, to restore it to its pristine state by stopping all farming and human impacts. I like this idea too in principle, and in later chapters I share how animals are essential for rewilding projects and why. But rewilding becomes less practical as the human population grows and exceeds our planet's resources. One of the many challenges of rewilding is that landowners are usually penalized financially because their land can no longer be farmed or managed. Wealthy individuals purchasing farmland for rewilding may raise land prices and make it harder for farming families to stay in business. Indigenous people could even be forced off their land. Could a new flavor of colonization appear if land is gobbled up by the super-rich in developing nations?

Already, wealthy elites have gained access to the ancestral lands of the Maasai to set aside for trophy-hunting activities. In 2017, the Arusha regional administration in Tanzania announced that a 1,500 $km^2$ area east of the Serengeti National Park had been leased to the Otterlo Business Corporation, a safari company with links to the United Arab Emirates. Supposedly for conservation reasons, several tribes of Maasai cattle herders who have coexisted with wildlife there for 400 years were evicted from their ancestral range in a poorly compensated land grab, according to an article in *The Atlantic*.[20]

The author, Stephanie McCrummen, interviewed many affected Maasai people, and she learned that in June 2022, the Tanzanian government forcibly evicted 70,000 Maasai herders from their land and locked up many who had protested in overcrowded prisons. Many others fled to Kenya. This is a missed opportunity for everyone on the planet, as 540 Maasai cattle and sheep herders have undergone training to use holistic management for ecosystem restoration at the Mara Training Centre in Kenya. If they can restore their ancestral lands as they intend to, it will benefit the entire planet and everyone on it, not just the super-rich. After you finish this book, you will have a deeper understanding of why the Maasai should be supported as Indigenous caretakers of their land, especially since they traditionally don't kill wildlife or cut down trees because their livestock provide almost everything they need.

**20** Stephanie McCrummen, "'This Will Finish Us': How Gulf Princes, the Safari Industry and Conservation Groups Are Displacing the Maasai from the Last of Their Serengeti Homeland," *The Atlantic*, April 8, 2024.

## Supporting Wildness at Raindance Farm

My holistic journey started in 2006 when I read about the land-management tools for restoring landscapes described by Allan Savory. Wildlife managers can use rest from fishing to allow ocean plants and animals to grow in numbers. Technology like earthworks is available, as is fire, although their use is usually harmful to ecology. The other tools are human creativity and holistic planned grazing, which I realized I could use to my advantage.

I had previously assumed that holistic planned grazing, which was growing more popular on beef cattle ranches, would not suit dairy farms like mine because dairy cows require higher-quality pasture. I changed my mind when Greg Judy, a rancher and holistic management practitioner, gave a talk about his beef farm in Missouri at a meeting for grass farmers in my state, New York. What impressed me were Greg's photos of the diverse wild pasture plants in his fields—the result of his practice of frequently moving his cows from place to place to mimic the behavior of wildlife herds. Greg's pastures were lush; I saw big, fat, green leaves of dandelion and plantain that looked like lettuce in a mesclun mix, along with red and white clover flowers in a bed of dark green grass. I knew my cows would love to graze pastures so green and lush. The flowers reminded me of the mountain meadows in the Swiss Alps, where milk for some of the world's best cheese is produced. I thought that if my cows had more herbs like these in their diets, they would produce better-tasting, healthier milk. The cheesemongers I had recently consulted in New York City had told me taste would be the key to success for the cheese we were thinking of making.

I read Allan Savory's book, *Holistic Management, A Commonsense Revolution to Restore Our Environment,* to find out how to generate the profitable results Greg was achieving. I started to see myself as a regenerative farmer instead of a tractor driver watching my soil blow away behind my tillage equipment. I began cultivating a deeper understanding of how nature works. Once I started learning about managing natural resources holistically and thought about how it could work for me, I never looked back. My family and I envisioned a thriving, diverse landscape that would provide us and our descendants with clean water and abundant food. Then my children, my employees, and I began taking actions that would help achieve that landscape. We began to manage our cows more thoughtfully, planning their grazing locations carefully and deliberately placing them in the right place at the right time to promote better plant diversity. At the same time, we began to consider additional measures we could introduce on the farm to protect the existing wildlife. This was not the same as the "rotational grazing" I had practiced for the previous 10 years, and the results were much better.

Within two years of letting our pasture rest a little longer between grazing events, mobbing our cattle to move more often but in a more condensed herd, and monitoring the results, we noticed obvious changes. More light purple flowers of red clover, yellow flowers of dandelion and bird's-foot trefoil, and blue/green leaves of sweet-tasting timothy grass appeared in our pasture. Wild perennial flowering plants sprouted up. We then knew we were on the right track for a healthier farm and the outcomes we wanted, including great-tasting dairy products.

After we started this new regime, our farm, which had earned the name "Raindance Farm" because grass grew very slowly in hot summer weather, began to respond differently to the dry summer weather. Dry spells had been a challenge for us, as we waited for rainfall to water our fields after significant investments of time and resources to reseed grass. We could not return the seed that had not germinated, nor recoup the money and fuel spent preparing the land for planting. We could only face the same bills with less income. When weather forecasters bubbled with enthusiasm as they shared the prospect of more sunny weather to come, we lost hope as we watched our pastures shrivel up. We had to scramble to buy hay to feed our cows during droughts, knowing the cost would inevitably rise as demand for it went up. But the new, holistic approach brought us hope of alleviating all that.

After three years using holistic planned grazing, our pastures stayed green, no longer drying out to a straw-colored, lifeless carpet during hot, dry Central New York summers. This meant less work and expense because we didn't have to buy hay bales and transport them to the fields to feed the cows. We were drought-proofing our farm and building resilience. With lower expenses in the summer growing season, I watched with relief as the milk truck pulled away every other day with our healthy, more economically produced organic milk to sell. We took care of the cows the best we could, and they put a roof over our heads. It was a symbiotic relationship.

## Sun Cheese

The short chain of production from the sun earned our cheese the name Sun Cheese. First our pasture plants harvested solar energy using photosynthesis and converted it into carbohydrates to grow leaves. Our cows grazed on the plants, turning this green chemical energy into meat and milk. The third step was making cheese in our solar-powered farmstead cheese plant. Our local Cooperstown Farmers' Market customers were able to buy Sun Cheese that had traveled just eight miles in its entire production cycle. Sun Cheese seemed to be a fitting name for it.

Our soil became more absorbent as its carbon content increased rapidly during the first six years of holistic planned grazing. Some plants in our pasture grew roots deep (to at least 800 mm) underground and sourced moisture from deeper in the soil profile, while taller plants shaded the soil and protected it from drying winds. When it rained, even in thunderstorms, our fields absorbed the rain more quickly, so little ran off. We knew we were encouraging soil biology to restore a more functional water cycle. When remnants of Hurricane Irene blew in torrential rain in 2011, the streams on our farm flowed crystal clear, only to join the brown torrent pouring off neighboring cornfields.

People can live without oil or gold, but without water, all life returns to dust. What kind of vegetation is covering the land near you? Do big rain events soak into the soil to recharge your aquifers and feed clear springs days or weeks later, like they should? Or does the rain run off rapidly in sheets to turn waterways brown and cause flooding somewhere downstream? This matters because worldwide access to fresh, clean drinking water is becoming increasingly difficult. In 2022, a total of 2.2 billion people lacked access to safely managed drinking water, according to the United Nations Joint Monitoring Programme for Water Supply, Sanitation and Hygiene.[21]

At Raindance Farm, pasture or hay left uneaten by the cows was no longer called "waste." We managed the cows so they could pay this carbon-containing plant biomass forward by trampling it down to the soil surface like the large herds of bison that once migrated across America must have done because of their sheer numbers. I had noticed that unless uneaten grass was trampled down to contact the soil surface, the earthworms couldn't reach it and incorporate it into the soil to feed other biology. We became confident that the carbon and minerals contained in the plants would pass through the many life forms in the soil until they helped fertilize more grass once again. With the help of our cows, we were jumpstarting the carbon and mineral cycle on our land and making it naturally more productive.

---

**21** UNICEF and WHO, "Joint Monitoring Program for Water Supply, Sanitation and Hygiene (JMP)—Progress on Household Drinking Water, Sanitation and Hygiene 2000–2022: Special Focus on Gender" (2023): 1.

Ancient megafauna, on savannas that were yet to be limited by human hunting, helped turbocharge nutrient cycling in the same manner. Megafauna are animals over 45 kg (100 lb) in weight, a definition that fits over 150 species, extinct and extant, including humans. It's the grazers that concern us here. They lifted productivity far higher than the few grazing species left today can. Some grazed ecosystems still enjoy some degree of faster nutrient cycling, thanks to grazing livestock. The presence of animals in many instances helps draw down more carbon from the atmosphere and store it in plant biomass.[22] Productive agroecosystems are going to feed humanity going forward so it is important for everyone to know how to achieve them with less fertilizer made from fossil fuels.

## Monitoring Results

A big part of holistic management involves monitoring the results we get from different management changes. I carefully looked at our finances to see the effects on our bottom line because I had employees and bills to pay. Monitoring our biodiversity was eye-opening. I learned to identify the new plants emerging and often got down on my knees to show my children (and anyone who would listen) what the earthworms and beetles were up to as they harvested the leaf litter our cows fed to them and drew it down into the soil to feed other species.

When we welcomed student groups from the local primary school to the farm, a highlight of the field trips was competing to see which team could find the most insects and worms in the pasture. Schoolchildren were invited to step out of their comfort zone and into the paddocks to rummage through week-old cow dung pats with the tools we provided. If we had used pharmaceutical dewormers, our cow manure pats would have been devoid of life, but instead they were teeming with arthropods. I would have loved to be a fly on the wall as they told their parents what transpired on their field trip! Most young people today don't get a chance to experience nature up close and get authentically dirty. This is extremely dangerous because the more disconnected people are from nature, the harder it is for them to honor and protect the very thing that provides all we need to live.

As plant diversity ramped up at Raindance Farm, flowers and seeds attracted insects, birds, and wildlife. I never expected to witness the profound domino effect on plant and wildlife diversity and abundance that resulted from honoring life and partnering with it. By grazing our cows on one small area of the farm at

---

22 Matteo Rizzuto et al., "Rewiring the Carbon Cycle: A Theoretical Framework for Animal-Driven Ecosystem Carbon Sequestration," *Journal of Geophysical Research: Biogeosciences* 129, no. 4 (2024): 10, https://doi.org/10.1029/2024JG008026.

a time, we allowed 98 percent of the farm to be untouched by livestock on any given day. Since we were organic, no poisons harmed the land, either. This allowed many plants, wild animals, birds, and insects to flourish. Ground-nesting birds got busy building nests and raising young. Crickets and cicadas sang by day and were chased by wild turkeys. Lightning bugs sparkled at night as white-tailed deer grazed in their midst.

We were beginning to make some progress in reversing the decline of wildlife that has been associated with agriculture for the past 10,000 years. I saw Eastern bluebirds darting from fence post to fence post on Raindance Farm in 2013. I had never seen the sky-blue flash of the Eastern bluebird before in the half-century that I had lived in New York, even though they are the official state bird. I suggested to my friends attending my 50th birthday party that they could bring me a bluebird nesting box built from scrap wood in lieu of a store-bought gift because I didn't need any more stuff for myself. I promptly installed the boxes on fence posts in pastures far away from my house cats. That summer, a bluebird family moved into one, and I witnessed the vibrant blue flashes every time I passed when leading my cows out to pasture.

Bobolinks, a nationally endangered prairie ground-nesting bird, became so abundant that their melodious song could be heard in every pasture. Build a habitat, and the former dwellers of the prairie will come. There were so many that I was amazed I had never noticed any before we started holistic planned grazing. The male bobolinks, with their striking yellow, black, and white uniform, always managed to grab our attention as they warbled while clinging to tall stalks of grass. They provided a whole new playlist for birdsong on the farm. I wouldn't dream of plugging my ears with earbuds for fear of missing out on any new additions to our farm orchestra. We noticed our cows kindly didn't step on the nests in the pastures they grazed. This makes sense since they evolved with ground-nesting birds over millions of years.

When we took our cows out to pasture after milking early on misty summer mornings, we saw thousands of spiderwebs draped from the tallest fronds of grass like diamond-studded dream catchers, dripping with dew. One might think our vegetable garden would be at risk with so many insects and birds, but our organic garden and fruit trees benefited immensely from the abundance and variety of life surrounding them (**Figure 2.1**). We saw little damage from insects and grew sweet potatoes the size of footballs, thanks to the composted cow dung that was used as fertilizer. Not a single apple from the 40 apple trees dotted around the farm had coddling moth caterpillars eating them, thanks to our bird and spider allies.

**Figure. 2.1 My daughter Keira Higgins in our backyard veggie garden at Raindance Farm.**

Photo by Siobhan Griffin (2015).

We didn't need artificial fertilizers or sprays, even though the commercial apple orchard five miles away soaked their trees in several poisonous sprays a year. Although occasionally a slug beat us to our big, juicy strawberries, we had more than enough to enjoy. The circle of life was reconnecting its severed arcs and becoming a stronger, better-functioning whole.

The diversity of insects (what pesticide companies call "pests") made our agroecosystem stronger and more resilient. When colony collapse killed our honeybees, I began to worry about future food production since honeybees pollinate so many fruit, vegetable, and nut crops. But soon after we found the dead hives, I was fixing a fence among the wildflowers (the ones chemical companies call "weeds") that grew along most of my fence lines. As I knelt to tighten a low wire, I realized I was surrounded by native pollinating wasps and wild bees that were buzzing around my ears on their journey from flower to flower. We worked side by side all morning like children in parallel play. There were as many different species of pollinators as there seemed to be kinds of flowers in our pastures, making my farm more resilient—future-proofing it.

By learning how to increase plant diversity and apply megafaunal impact to cycle nutrients faster, we grew more plant biomass at the foundation of the food chain to feed our cows as well as snails, slugs, insects, salamanders, deer, ducks, voles, Cooper's hawks, sparrow hawks, bald eagles, owls, coyotes, bobcats, fishers, all the way up to the top predator: humans. This was in stark contrast to monocultures of corn or vegetable crops grown with chemicals that kill everything else, even the life in the soil. You don't need a farm to enjoy the satisfaction of restoring nature where you live. Birds and insects have wings and can find you. Create a habitat of diverse plants instead of a lawn or paved area near you and see who comes to visit. Support local farmers who have created healthy wildlife habitat on their land by being their loyal customer.

We knew our soil biology enjoyed beneficial cascading effects because our soil's organic matter had increased enough to sequester at least 9,000 tonnes of $CO_2$ equivalent on our 89 grazed hectares over 10 years. Learning how to produce food with a smaller carbon footprint should be important to everyone. No doubt there are other effective approaches to rapidly build soil carbon, but farmers and ranchers considering switching to holistic planned grazing can rest assured that feeding overmature brown pasture to livestock is not required to build significant soil organic matter in temperate climates. What we did was effective because we were focused on productivity and growing green leaves as fast as we could. This went against the advice of many self-proclaimed regenerative farming experts who think if a little more rest for pastures is good, then a lot of rest is even better. This may be the case in semiarid regions but it wasn't the case for us. Animal performance and feeding the soil were our priority, thanks to coaching by a holistic management educator named Ian Mitchell-Innes.

We were not the only farmers seeing soil carbon rise with more intentional grazing management. Based on differences in outcomes on regeneratively grazed farms versus conventional farms, a 2020 study, conducted by the State University of New York at Albany, determined that multispecies adaptive grazing could reduce the carbon footprint of livestock farming in the US by 35 percent if widely adopted. The author of the study, Professor Emeritus of Biology Gary S. Kleppel, observed the irony that growing food conventionally to sustain our human population is destroying the ecosystems required for life. He concludes, "Livestock production represents a tool, not only for providing calories and protein to our species over the next half century, but for ensuring the livability of our planet to humans and myriad other species well into the future." [23]

**23** Gary S. Kleppel, "Do Differences in Livestock Management Practices Influence Environmental Impacts?," *Frontiers in Sustainable Food Systems*, 4 (2020): Conclusion, para.3, https://doi.org/10.3389/fsufs.2020.00141.

# Chapter 3

# How Pristine Ecosystems Cooled Planet Earth

As a species, we use tools that change the biosphere so severely that we change the global climate, and we were already doing so thousands of years before the advent of agriculture. Understanding the relationship between the life that makes up the biosphere and our climate holds the key identifying to the most powerful tools to address climate change—land and fisheries management. For instance, when human hunting pressure decimated mammoth, horse, and bison populations around 13,800 years ago, dwarf birch trees (*Betula nana*) outcompeted grasses on the mammoth steppe. Because trees and shrubs are darker than grasslands, their widespread expansion over a millennium reduced the ability of the land to reflect the sun's energy out to space. This reduction in reflectivity (albedo) may have caused enough localized warming to raise regional temperatures by up to 1 degree Celsius.[24] This pushes the onset of the Anthropocene, the era in which humans affected global climate, much farther back than scientists had initially considered. But how else does biology affect global climate? And can you use this knowledge to manage nature where you live in order to influence the climate?

Over the last 100 years or so, humans have mined and pumped carbon from the lithosphere deep underground at rapidly increasing rates. The result is many symptoms of ecological overshoot, including a steep spike in the $CO_2$ concentration in the atmosphere. Alarming evidence of this spike in $CO_2$ has been revealed by ice core samples from Vostok Station, a Russian research facility in Antarctica (**Figure 3.1**). Over the past 800,000 years, when atmospheric $CO_2$ rose so did temperature. These metrics declined together, too. So, we must ask ourselves: what role does nature play, if any, in cooling the Earth's climate?

**24** Christopher E. Doughty et al., "Biophysical Feedbacks Between the Pleistocene Megafauna Extinction and Climate: The First Human-Induced Global Warming?," *Geophysical Research Letters* 37, no. 15 (2010): fig. 2, https://doi. org/10.1029/2010GL043985.

The strength of the sun's energy, or solar insolation, at the northern hemisphere, along with macro-oceanic circulation, has long been assumed to control the fluctuations in temperatures seen during the many glacial periods during the ice age Earth has been experiencing for the last two and a half million years. But after each of the interglacial warming periods between cold glacial periods over the past 800,000 years, ice core data show declining trends in atmospheric $CO_2$ and global temperature, as demonstrated in **Figure 3.1**. The decline of these two metrics, almost in lockstep, result from complex feedback loops between the biosphere, oceans, land, and the atmosphere.

Glaciers grew as orbital cycles weakened the sun's energy arriving at the northern hemisphere. Through many feedback loops, the planet cooled further. For instance, cold ocean water absorbs more atmospheric $CO_2$ than warm ocean water does, producing another cooling feedback effect. Nevertheless, scientists found a distinct positive correlation (.71 for $CO_2$ and .73 for methane) between these greenhouse gases and global temperature over the past 400,000 years and think they may have contributed to half of the temperature changes between ice ages and the warm interglacial periods.[25] In other words, greenhouse gas levels, not just solar insolation, also affect global temperature. This is especially worrisome if you look at how atmospheric levels of $CO_2$ have exploded to 432 ppm in the last two centuries, just a blip in geological time (**Figure 3.1**).

## Megafauna, Soil Carbon, and Global Climate

As the planet cools toward subsequent ice ages after warm interglacial periods, increasing glacial ice lowers sea levels. As more land emerges from the sea, it is colonized by green photosynthesizing plants and algae that lift the carbon-containing biomass of life and initiate soil- and peat-building over vast areas. This carbon drawdown from the atmosphere via photosynthesis is a cooling force of nature. Some carbon, especially that which is associated with minerals in soils and ocean sediments, can be stored for hundreds or thousands of years.

Gregory Retallack is a soil scientist who is intrigued by mollisols. These are carbon-rich, dark soils that form under temperate grasslands. Their crumbly structure acts as a sponge for carbon and water and helps grasslands sequester far more carbon than the woody scrub and forest ecosystems they replaced. Historically, deep grassland roots allowed organic matter to build up to high

25 J. R. Petit et al., "Climate and Atmospheric History of the Past 420,000 Years from the Vostok Ice Core, Antarctica," *Nature* 399, no. 6735 (1999): 434, https://doi.org/10.1038/20859.

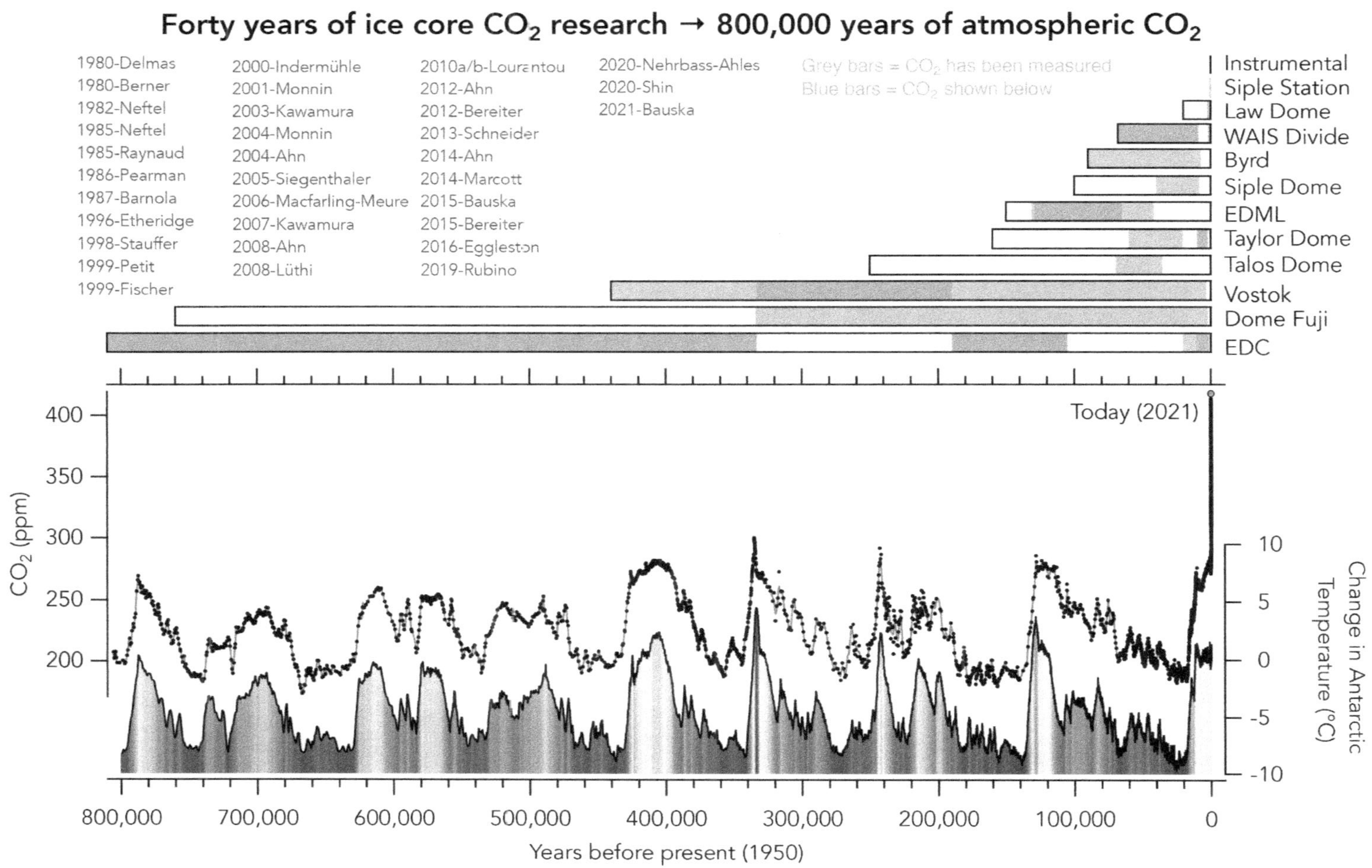


**Figure 3.1. $CO_2$ and temperature over the last 800,000 years.** Reproduced from Bauska, 2025.

levels over millennia in places with these soils, such as North American prairies, South American Pampas, northern Australia, and parts of China and Ukraine.[26] It is no coincidence that these places are some of the most productive grain-growing regions on the planet. Retallack points out that after every interglacial spike in global temperature, atmospheric $CO_2$ and global temperature gradually decline because of the interplay of climate-forcing factors at work, including the Earth's orbit. He suggests, however, that the contribution of the soil-building activities of biology on grasslands and savannas to cool the planet is often underestimated. Retallack believes that the global evolution and expansion of tall grasslands over the past 40 million years may have resulted in the many periods of glacial expansion we call the ice age.[27] In his 2022 book *Soil Grown Tall* he writes,

> Restoring grasslands could help sequester atmospheric carbon dioxide in our current greenhouse crisis. Just as carbon storage in expanding grassland soils ushered in a global ice age, carbon storage and water capacity in soils can be improved by agricultural practices, such as holistic planned (or cell) grazing used with success by Allan Savory and Richard Teague, among others.[28]

Luciano Prates and Ivan S. Perez studied historical ecosystem changes in South America. They observed that pampas and cool steppe grasslands expanded, and rising megafaunal numbers peaked, with the Antarctic Cold Reversal between 14,700 and 13,000 years ago. This cooler phase in climate in the southern hemisphere ended with the appearance of fishtail projectile points used by human hunters of large mammals. According to Prates and Perez, "The indirect effect of humans on the ecological web probably generated the conditions resulting in the massive collapse of the megafaunal community."[29] Megafaunal collapse around 12,500 years ago heralded a return to long-term warming in the southern hemisphere. These findings suggest that humans were involved in the extinction of South American megafauna as well as the ecological changes to the pampas and Patagonian ecosystems at the time, with knock-on effects on climate.

---

**26** Gregory J. Retallack, Soil Grown Tall (Springer International Publishing, 2022): 85, https://doi.org/10.1007/978-3-030-88739-1.

**27** Gregory J. Retallack, "Global Cooling by Grassland Soils of the Geological Past and Near Future," *Annual Review of Earth and Planetary Sciences* 41, no. 1 (2013): 70, https://doi.org/10.1146/annurev-earth-050212-124001.

**28** Retallack, *Soil Grown Tall*, 87.

**29** Luciano Prates and S. Ivan Perez, "Late Pleistocene South American Megafaunal Extinctions Associated with Rise of Fishtail Points and Human Population," *Nature Communications* 12, no. 1 (2021): 9, https://doi.org/10.1038/s41467-021-22506-4.

## Water Vapor, the Most Potent Greenhouse Gas

Water vapor is the greenhouse gas that controls 95 percent of global heat dynamics, but it is often ignored in climate models because many people believe its dominance is too big to be influenced by human activity. Australian microbiologist Walter Jehne suggests they should think again. He is working beyond his retirement to educate farmers and policymakers on the importance of restoring water cycles to reduce climate deviation from the temperatures humans have enjoyed as we evolved over the last two million years.

Jehne uses the concept of the "soil carbon sponge" to describe the porous, well-aggregated structure and high organic content of good-quality soil. Natural terrestrial ecosystems sequester carbon on an ongoing basis by growing topsoil and peat.[30] Jehne believes that rebuilding the lost carbon sponge across the vast areas of the Earth where it has been destroyed by overgrazing and unsustainable arable agriculture is a powerful climate mitigation action. The benefit to climate will not be merely from the atmospheric $CO_2$ reduction, but from the ability of carbon-rich soils to absorb water and enhance the water cycle.[31]

If you have ever dug a hole in a forest or grassland to plant a tree and found soft, crumbly, earthy-smelling soil, this is the soil carbon sponge. The interaction of life within and above healthy soil creates topsoil with a porous, light structure. Soil humus, the dark, sticky, absorbent, earthy-smelling substance that defines healthy soil, is built with glues created by microbes, including fungi. These microbes use carbon building blocks that plants exude into the soil in both liquid and solid form as the raw ingredient for the humus building process. Thus, soil is not just ground-up rock and minerals, but is also made by life that's powered by sunlight.

The volume of healthy soil should be 50 percent air to grow plants well. Vermonter Didi Pershouse is dedicated to teaching people about how biology works. She illustrates the soil carbon sponge to farming and conservation groups in an interesting way. She puts the dry ingredients for a loaf of bread on a plate and pours water on top of the plain flour and dormant yeast. It immediately runs off without soaking in. This mirrors how rain runs off quickly without soaking into over-tilled soil containing little life. Then, she takes a slice of bread, baked with the help of living yeast, and puts it on the plate. Water poured on it immediately soaks

30 Retallack, "Global Cooling," 85.

31 Walter Jehne, "Carbon Capture and Storage: The Biological Potential," *Proceedings of the Royal Society of Victoria* 126, no. 2 (2014): 18, https://doi.org/10.1071/RS14018.

in as it fills pore spaces, displacing air. The baked bread represents healthy soil with air spaces. These spaces are built by creatures living underground, symbionts in the web of life. When it rains, these spaces become reservoirs for water that the soil can hold onto like a sponge until plants need it, extending green growing seasons into dryer weather. Jehne calls this "longevity of green," and it plays an important role in healthy water cycles that cool the planet.

Green plants help create rain by pulling soil moisture from the carbon sponge storage area in healthy soil and releasing it into the air as water vapor through holes in their leaves called stomata. This process, called transpiration, provides water vapor destined to become rain and is especially important in the interior of continents far from oceans, where plant transpiration recycles water into rain repeatedly from the original oceanic sources of moisture. Eighty percent of the rain falling in China, for example, is recycled rainwater produced from water vapor released by vegetation in Europe.[32]

Some trees help rain droplets form in clouds by releasing symbiotic bacteria such as *Enterobacteriaceae* (formerly known as *Aerobacter*) from their leaf stomata.[33] Other ice-nucleating bacteria, including *pseudomonas*, as well as fungi have been identified in rainwater samples.[34] These float into the air currents to act as seeds for raindrops to coalesce upon.

We can witness life creating rain today in the Amazon tropical rainforest. The same molecule of water that enters the Amazon rainforest from melting glaciers high in the Andes Mountains may be recycled by a relay team of living organisms several times before exiting the Amazon for the last time as river water. Forest biomes like the Amazon rainforest even create low atmospheric pressure above themselves due to the condensation of the water vapor they release. This draws moisture inland from adjacent oceans in what has been coined "the biotic pump."[35] The net result is a shorter dry season and higher ecosystem productivity.

The power of the carbon sponge to positively impact water cycles and thus planet hydrology, Jehne points out, has a much stronger impact on climate than the related carbon cycle. Transpiration by plants is one of the most important

---

**32** Rudi J. van der Ent et al., "Origin and Fate of Atmospheric Moisture over Continents," *Water Resources Research* 46, no. 9 (2010):1, https://doi.org/10.1029/2010WR009127.

**33** Walter Jehne, "Restoring Water Cycles to Naturally Cool Climate," remarks to the Soil Carbon Sponge seminar, video, November 30, 2016, 1:20–1:23, https://soilcarboncoalition.org/walter-jehne-water/.

**34** K. C. Failor et al., "Ice Nucleation Active Bacteria in Precipitation Are Genetically Diverse and Nucleate Ice by Employing Different Mechanisms," *ISME Journal* 11, no. 12 (2017): 2740–53, https://doi.org/10.1038/ismej.2017.124.

**35** Anastassia M. Makarieva and Victor G. Gorshkov, "The Biotic Pump: Condensation, Atmospheric Dynamics and Climate," *International Journal of Water* 5, no. 4 (2010): 367, https://www.inderscienceonline.com/doi/abs/10.1504/IJW.2010.038729.

sources of water vapor in the upper atmosphere. If the water vapor turns into clouds, it reflects solar energy into space and helps cool the planet by higher albedo as well. In addition, when this water vapor finally condenses into clouds and rain, energy is released high in the atmosphere, where it can radiate back out to space more easily. This cooling effect of our planetary hydrology uses the same process that occurs when an air conditioner releases heat to the outdoors from the phase change of coolant turning from a gas into a liquid. The more people can build healthy soil, protect natural or diverse forests, and green our planet, the more we can restore the water cycles essential for a healthy planet and help cool her rising fever.

## Savannas Stalled Ice Age Cycles

The impact of ongoing atmospheric carbon removal by grasslands could be the missing piece to a longtime puzzle for paleoclimate scientists. During the mid-Pleistocene transition, which occurred about a million years ago, ice age cycles switched from every 41,000 years to about 100,000-year cycles. The extremes of sea level change between interglacial warming periods and the coldest depths of the next ice age also increased, from a difference of 60 m to a difference of 120 meters, indicating that the climate cycle reorganization was a significant change.

Scientists have debated why this pattern emerged, even though the solar radiation patterns (also known as Milankovitch cycles) due to the Earth's orbit did not change much in that period. These cycles align the Earth's northern hemisphere to face the sun more directly or bring the Earth physically closer to the sun, which causes the melting of glaciers during the warm periods between glacial times (i.e., interglacial periods). Or the combination of the Earth's orbit, tilt of its axis, and distance from the sun can produce the reverse effect if the northern hemisphere is tilted away and is farther from the sun, which causes glaciers to grow.

One explanation for the extra cooling that prevented a warm interglacial period every 41,000 years is regolith removal. Regolith is loose soil, rocks, and/or gravel that allows glaciers to slide more easily over bedrock. When regolith sits below glaciers, the ice remains relatively thin because the glacier moves faster toward the ocean or lower latitudes, where it melts. After the glaciers of successive ice ages pushed away the glacial till underneath them, hard bedrock was exposed. This was stickier and slowed the glacier's movement, so the glaciers piled up higher in that location. The thicker ice was then slower to melt, and some scientists think this kept the planet cooler for longer and helped to prevent a warming period every 41,000 years or so, as the Earth's orbital cycles would have predicted.

But computer modeling shows that regolith removal alone is not enough to prevent 41,000-year ice age cycles.[36] Another significant driver for long-term atmospheric $CO_2$ reduction was required. I posit that the ability of life to store carbon long-term in soils, peats, and ocean sediments may have played this role. The actions of soil-building biology on grasslands that result in improving the soil's carbon levels and water storage capacity, and the cooling effect of rain clouds produced by forests, could have enhanced planet-cooling hydrology and may have been the mechanisms missing in the climate models.

When we built our cheese cave at Raindance Farm, we needed to bury water lines to run to the facility, so we ran them through one of our pastures. Because air temperatures can drop to –34°C in winter, we dug down four ft (just over 1 m). I was thrilled to see plant roots had penetrated down to the bottom of the trench, and the soil had a crumbly texture that indicated soil life was thriving. Our carbon sponge was one meter thick.

In the top 1 meter of grassland soils on the planet today, 47 percent of the carbon is stored deeper than 30 cm; this carbon tends to be more stable and is much older. The total of the world's soil carbon in the top 1 m is massive, totaling 1,500 gigatonnes,[37] not including permafrost.[38] There is more carbon in soil than exists in all the above-ground plants and animals on the planet. This means land management practices that result in soil-building with deep-rooted plants have massive potential to address today's increasing climate volatility.

## Soil Washing Away

Holistic management is a way to make decisions that considers the things we can manage: economies, the environment, and people. I co-managed a confinement dairy farm in Connecticut in 1990 and 1991 with my husband at the time. The adult cows were housed indoors throughout the year, and we grew corn and hay to make silage that we fed to them. The owner enjoyed the high yields that growing 120 hectares of conventional corn provided and, with a reductionist viewpoint, insisted we grow as much corn as possible. One field that had been growing corn continuously for several years was on a slope. One summer day during a thunderstorm, tonnes of soil washed off the field and into the road ditches.

36 M. Willeit et al., "Mid-Pleistocene Transition in Glacial Cycles Explained by Declining $CO_2$ and Regolith Removal," *Science Advances* 5, no. 4 (2019): Disc., para. 1, https://doi.org/10.1126/sciadv.aav7337.

37 1 gigatonne is a unit of mass equal to one billion metric tonnes.

38 Jérôme Balesdent et al., "Atmosphere–Soil Carbon Transfer as a Function of Soil Depth," *Nature* 559, no. 7715 (2018): 599, https://doi.org/10.1038/s41586-018-0328-3.

Even though the corn was 1 m tall there was a lot of bare soil between the rows, with no plant roots to hold the soil in place.

We took the owner down to see the field and suggested to him that this field might be better suited for a hay crop, which would reduce soil erosion. The owner refused to let us change the management of that field because corn would yield more tonnage than hay. Considering only the financial bottom line can misguide management if it is not put into perspective with potential future impacts that can weaken a business in the long run. It is worth considering the social implications of "bottom-line" decisions, too. In this case, the owner lost his co-managers because we did not want to be involved in such unsustainable farming practices. The experience made me realize that confinement feeding of dairy cows was not how I wanted to farm.

The magnitude of the carbon-rich topsoil lost on two-thirds of the world's land area in grassland, steppe, or savanna since the last interglacial warm period is hard for humans to appreciate, coming from the context of our present, degraded human habitats. The carbon loss from cropping and grazing in the last 10,000 years is estimated at 116 billion tonnes of carbon, but scientists warn this could be far higher.[39] Recently, researchers measured the erosion caused by farming on the American prairie. Twenty prairie remnants and adjacent agricultural fields in the Corn Belt were studied. Since European methods of plowing were introduced, topsoil has been shrinking at a median erosion rate of 1.9 mm per year, more than twice the sustainable rate.[40] When soil loss is multiplied by billions of hectares it amounts to a very significant release of carbon into the atmosphere and oceans. Global agricultural soil loss is estimated to be 75 billion tonnes per year. Worldwide, two billion hectares of cropland have been degraded and abandoned by humans since farming began. This is more than the 1.5 billion hectares crops are grown on today.[41] Therein lies the opportunity humanity needs right now—land to clothe with a protective green cover of life.

---

**39** Jonathan Sanderman et al., "Soil Carbon Debt of 12,000 Years of Human Land Use," *Proceedings of the National Academy of Sciences of the United States of America* 114, no. 36 (2017): 9578, https://doi.org/10.1073/pnas.1706103114.

**40** Evan A. Thaler et al., "Rates of Historical Anthropogenic Soil Erosion in the Midwestern United States," *Earth's Future* 10, no. 3 (2022): 9–10, https://doi.org/10.1029/2021EF002396.

**41** David Pimentel and Michael Burgess, "Soil Erosion Threatens Food Production," *Agriculture* 3, no. 3 (2013): 47–48, https:// doi.org/10.3390/agriculture3030443.

## Methane-Producing Megafauna Cooled the Climate

During the cooling phases of the ice age, more land became available for grazing animals as sea levels dropped. As savanna and grassland ecosystems spread, the concentration of methane in the atmosphere trends downward. This is counterintuitive because an increased land area for animals would support more animals, which would result in more climate warming methane being produced. However, natural recycling of carbon prevented enteric methane from building up in the atmosphere. Despite the abundance of large grazers at a peak in their numbers 23,000 years ago, methane levels in the air were just under one-third of what they are today.

Nature has several pathways for recycling methane. It was recently discovered that methanotrophic bacteria living in tree bark use methane as an energy source and break significant quantities of it down into the less potent greenhouse gas $CO_2$.[42] In nature's brilliant design, the lightly wooded landscapes that many megafauna were embedded in filtered out the methane they released. Methanotrophs also live in forest and grassland soil.[43]

Although this would have partially reduced the buildup of atmospheric methane produced by ancient megafauna (as it does with wildlife and livestock on the savanna today), there appears to have been a much larger biological mechanism mitigating the warming potential of massive methane emissions from herbivores. Walter Jehne explains that in naturally functioning forest, grassland, or savanna ecosystems, as transpiration by plants pulls water from the soil and releases it as water vapor into the air, the water vapor provides sufficient hydroxyl radicals (-OH) to oxidize all the methane emitted by grazing ruminants and change it into $CO_2$.[44] This $CO_2$ will be promptly taken up again by plants in the process of photosynthesis if animals are present on the green ecosystems they evolved upon.

This is profoundly relevant to livestock systems today where cows, sheep, and pigs are removed from the green habitats they evolved in and are confined indoors, where no plants live to help mitigate enteric methane. This underscores the importance of managing livestock holistically in agroecosystems that mimic their natural environment. Here they can provide natural fertilizer that turbocharges plant growth that in turn draws down more $CO_2$ from the air.

---

**42** Vincent Gauci et al., "Global Atmospheric Methane Uptake by Upland Tree Woody Surfaces," *Nature* 631, no. 8022 (2024): 800, https://doi.org/10.1038/s41586-024-07592-w.

**43** Pascal A. Niklaus et al., "Plant Species Diversity Affects Soil–Atmosphere Fluxes of Methane and Nitrous Oxide," *Oecologia* 181, no. 3 (2016): 920, https://doi.org/10.1007/s00442-016-3611-8.

**44** Walter Jehne, "The Soil Carbon Sponge, Climate Solutions and Healthy Water Cycles with Walter Jehne," recorded by Biodiversity for a Livable Climate, presented on April 26, 2018 at Harvard University, 1:15:00–1:20:50, https://youtube/123y-7jDdbfY?si=v8f9ToXnf6AZTC6-.

The best way to encourage outdoor management of livestock in your foodshed, if you can't grow your own, is to seek out food grown on regenerative farms. Modern people have written off raising their own livestock even though suburban sprawl in many places offers plenty of green space to raise chickens, sheep, or goats. Without animals, people must do the work of cycling nutrients using lawn mowers and hedge trimmers. Instead of recycling nutrients back onto their properties, often homeowners send their leaves, prunings, and grass clippings away where their nutrients may be wasted. If sent to landfill, green waste can emit methane as it decays.

People in some places will need to change zoning laws along with attitudes to allow coexistence of people with the natural world again. When my daughter Keira was at university, she asked me to help her build a mini chicken tractor so she and her roommates could have fresh eggs at school. Chicken tractors are moveable pens that can be shifted around a big lawn or pasture to provide chickens with access to grazing and arthropods to eat. They have become very popular since regenerative farmer Joel Salatin wrote his book *Pastured Poultry Profits*, which was published a month before Keira was born in 1996.

One day, four burly City of Rochester policemen arrived at Keira's door to tell her she was partaking in criminal activities by having chickens on the lawn outside her student apartment. She didn't get fined because she agreed to find a different place to graze her chickens. I hope someday to see over-regulated communities embracing the use of grazing animals to keep fire breaks open and air-drying their laundry, ecologically friendly practices often forbidden in homeowner associations.

## Satellites Reveal Methane Super Emitters

The carbon in the methane of animal burps is sourced from $CO_2$ in the air that plants absorb via photosynthesis. In other words, enteric methane contains carbon that has already been circulating in the atmosphere. Fossil methane is a different story. This carbon has not been part of the current carbon cycle for millions of years. Many scientists report enteric methane from livestock farming as an environmental cost to meat and milk production. But according to the United Nations, "Fossil fuels—coal, oil and gas—are by far the largest contributor to global climate change, accounting for over 75 percent of global greenhouse gas emissions and nearly 90 percent of all carbon dioxide emissions."[45]

**45** United Nations, "Causes and Effects of Climate Change," 1, accessed August 15, 2024, https://www.un.org/en/climatechange/science/causes-effects-climate-change.

The gigatonnes of $CO_2$ and methane released from fossil fuel production are additional climate-warming gases drawn from ancient carbon cycles that have been trapped away deep underground in the lithosphere. Many human activities produce methane. However, just the amount of waste methane that is flared from global fossil fuel production is truly eye-watering. Flaring, or burning into the atmosphere, of methane at oil and gas production facilities is a common way to simply get rid of natural gas that comes up with the oil. It is usually wastefully burned to turn methane into the less-potent greenhouse gas $CO_2$. In 2024, the gas lost in flares burned 151 billion cubic meter of natural gas made up of methane and soot. According to the World Bank, an estimated US$63 billion worth of gas was wasted.[46]

Many wells are poorly flared or not flared at all and can become super emitters, as identified by NASA's methane-detecting satellite.[47] In 2024, the International Energy Agency estimated that "around 80 million tonnes of annual methane emissions from fossil fuels can be avoided through the deployment of known and existing technologies, often at low—or even negative—cost."[48] The wasted methane from fossil fuel production alone is approximately equal to the total global annual methane emissions from all ruminant livestock, which are estimated to be between 80 and 95 million tonnes.[49] Is the fossil fuel industry shifting blame onto cows for methane emissions to deflect scrutiny from its own activities? The more urgent issue is the fundamental problem with how livestock are raised today, separated from where their food is grown. However, with a better understanding of ecology, we can grow food with less dependence on fossil fuel inputs. Mastering high rates of natural ecosystem productivity in our agroecosystems is the key.

## The Path Forward

Past abundance and ecosystem productivity provide us with an important template of what is possible as we attempt to restore the cooling effect of healthier hydrology. Even if the use of all $CO_2$-emitting fossil fuels were halted permanently today, excess $CO_2$ from the oceans would burp back up into the atmosphere for

46 World Bank, *Global Gas Flaring Tracker Report* (2025): 15, https://thedocs.worldbank.org/en/doc/bd2432bbb0e514986f-382f61b14b2608-0400072025/original/Global-Gas-Flaring-Tracker-Report-July-2025.pdf.

47 Genevieve Plant et al., "Inefficient and Unlit Natural Gas Flares Both Emit Large Quantities of Methane," *Science* 377, no. 6614 (2022): 1566, https://doi.org/10.1126/science.abq0385.

48 IEA, "Global Methane Tracker 2024: Key Findings," https://www.iea.org/reports/global-methane-tracker-2024.

49 Wondimagegne Bekele et al., "Contemporary Methods of Measuring and Estimating Methane Emission from Ruminants," *Methane* 1, no. 2 (2022): 83, https://doi.org/10.3390/methane1020008.

centuries if not millennia. Luckily, nature can offset the greenhouse gas effect of excess $CO_2$ and continue to draw carbon down in the ongoing manner required. We must tap into these natural mechanisms and learn to amplify them if we want a livable planet going forward for millennia. Building a deeper soil carbon sponge is one commonsense approach to manage the water cycle better. For land stewards, deeper topsoil means more productive farms: a win-win of gigantic proportions.

I learned from experience that it is possible to produce beef and milk using lower carbon footprint methods, but the present food system subsidizes farmers to grow surplus grain, and this makes it hard for regenerative farmers to compete. Policy changes need to occur to make it easier for farmers to support more life on their farms. Citizens can help rewrite their own future by advocating for farm policy with less emphasis on corn and soybean subsidies.

A scientific investigation compared two systems to raise beef: the current maize crops grown for animal feed and ethanol on American tall-grass prairie, and a hypothetical grazed perennial grassland agroecosystem that could support much more wildlife. The paper's author, Randall Jackson, found that a shift to grazed agroecosystems would support the same quantity of beef that ranchers and feedlot operators produce today, and no new land would need to be converted to agriculture. The warming effect of enteric methane and all the other greenhouse gases produced by grass-fed beef could be offset if ranchers were to sequester just 1.63 tonnes of soil carbon per hectare per year. I believe this is doable because we achieved about double this rate of carbon sequestration with our dairy cows at Raindance Farm in our first 10 years using holistic planned grazing. Jackson concluded that the switch to a grazed agroecosystem "requires policies that incentivize production systems for the public good rather than profit of a few. And it will improve our quality of life overall."[50]

Reimagining livestock rearing where animals are returned to a naturally functioning savanna would allow methane-eating bacteria to help cycle the carbon in the enteric methane molecules. This agroecosystem would also support a plethora of wild species such as the bobolinks we saw nesting at Raindance Farm. It is industrial agriculture, not the animals, that must change. Livestock can be our allies and help us rewrite our future if they are managed in nature's image.

**50** Randall D. Jackson, "Grazed Perennial Grasslands Can Match Current Beef Production While Contributing to Climate Mitigation and Adaptation," *Agricultural & Environmental Letters* 7, no. 1 (2022): sect. 6, para. 2, https://doi.org/10.1002/ ael2.20059.

## Chapter 4

# Climate Stabilizing Life

For a few weeks every year, spring would be announced on our farm by the chorus of chirping calls from the spring peepers (*Pseudacris crucifer*) who perused our pond looking for mates. When it rained, the adjacent road was covered in regiments of these small but vocal frogs hopping around, with many succumbing to the light traffic in a slippery mess. Over subsequent days, they and a plethora of amphibians and turtles could be seen mating in an orgy of life that made the pond look like it was boiling! Was this annual display of nature's vitality a hint of what abundant life looked like in pristine ecosystems? What did pristine landscapes look like before people came? Our ancestors tried to tell us.

Over 30,000 years ago at what is now Chauvet-Pont d'Arc Cave in southern France, people recorded on cave walls the abundant game they saw lest we forget the magnitude of nature's pristine abundance (**Figure 4.1**). These included herds of woolly rhinos (*Coelodonta antiquitatis*); aurochs (*Bos primigenius*), the XL size ancestor to modern cows; woolly mammoth *(Mammuthus primigenius*); and giant steppe bison (*Bison priscus*). The green mosaic of vegetation resulting from animal impacts covered landscapes all around the planet and maintained water cycles. Rain fell in positive feedback loops to fill massive underground aquifers. The clouds that came from this meeting of life and the elements reflected solar radiation into space, cooling the global climate. The more productive the ecosystems were, the more powerful life was at building soil and water wealth and cooling regional climates.

## Old and New Extinctions

Our Paleolithic ancestors brought effective killing technology to the far reaches of the planet. By initiating megafaunal extinctions, humans destroyed the integrity of many ecosystems that previously produced the fossil water that is essential for growing food in semiarid regions today. These areas are now missing the protective and productive green cover of vegetation that once helped cool the

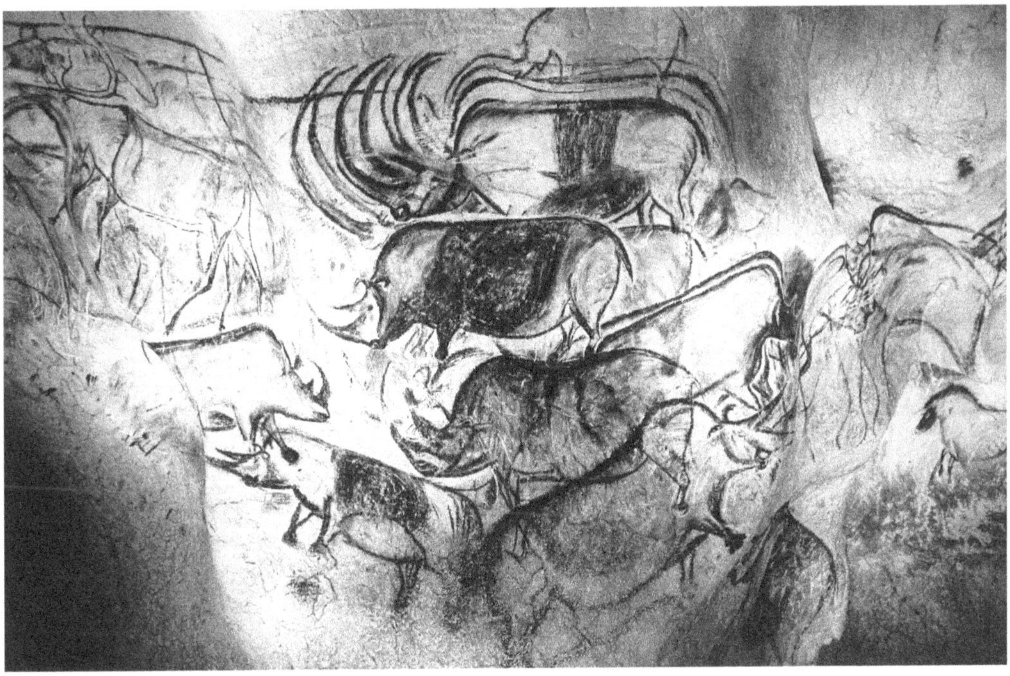

**Figure 4.1 "Rhinoceros and Lion Panels, End Chamber" Chauvet-Pont-d'Arc Cave (France).**

Photo by J. Clottes.

soil and replenish underground water. Denuded regions get ever hotter and drier as a result. But we can and we must reboot the water cycle by restoring the full complement of both flora and fauna so aquifers can fill once again.

Scientists often assume that climate change caused the demise of savanna on desertified lands, but observant scientists like Walter Jehne and David Wright suggest these highly evolved and diverse ecosystems were well-watered because of the presence of a protective green cloak of life.[51, 52] As these seasonally arid places lost their megafaunal gardeners and vegetation, the water cycle was interrupted enough that climatic tipping points were reached. Eventually, even the rains stopped falling. The desertification process initiated by humans can also be reversed by the way we manage our land! It is up to us to ensure our progeny will have access to the natural resources they need for survival, and fresh water and healthy soil are at the top of the list.

---

**51** Walter Jehne, "The Soil Carbon Sponge, Climate Solutions and Healthy Water Cycles with Walter Jehne," recorded by Biodiversity for a Livable Climate, presented on April 26, 2018, at Harvard University, 1:41:50–1:45:55, https://youtube/123y-7jDdbfY?si=v8f9ToXnf6AZTC6-.

**52** David K. Wright, "Humans as Agents in the Termination of the African Humid Period," *Frontiers in Earth Science* 5 (2017): Conclusion, para. 2, https://doi.org/10.3389/feart.2017.00004.

Today, the accelerating rate of a new mass extinction, the sixth, is eclipsing the speed of any previous defaunation. As it always has, cutting off entire genera and species from the tree of life damages the integrity of the whole biosphere and can destroy natural capital rapidly. This is why in 2020, the World Economic Forum rated biodiversity loss as "the third most impactful long-term risk factor" for global economies. The first was failure to mitigate and adapt to climate change.[53] Not enough effort has been invested in protecting important natural resources. Our oceans are overfished, and our rainforests and grasslands are being rapidly decimated because our cultures see non-human life as resources to be plundered instead of symbionts in our own web of life.

Our perception of what is pristine becomes more impoverished every generation in a process called "shifting baseline syndrome."[54] People do not realize that modern "pristine" ecosystems contain just a few tiny interlocking pieces of the giant jigsaw puzzle of the Earth's hard-won evolution.

## Evidence of Exceptional Abundance

*Homo sapiens* began migrating out of Africa and into Asia around 70,000 years ago and reached Western Europe at least 54,000 years ago.[55] There, Paleolithic people encountered lightly wooded grasslands and savannas on the plains and valley floors.[56] That land was already settled by Neanderthals (*Homo neanderthalensis*) who were fending off giant saber-toothed tigers (*Homotherium spp.*) and the oversized cave lion (*Panthera spelaea*) on landscapes teeming with life. For these early human tribes, highly skilled in hunting, the land provided an abundance of animals, fruit, and nuts. But what made this cornucopia of abundant life possible? It is crucial for people today to understand how these ecosystems became so productive. If human populations continue to grow, we will soon need to feed 10 billion people on planet Earth so regenerative farming must also be productive. Luckily, nature has already developed the productive templates we need to restore.

---

**53** WEF (World Economic Forum), *The Global Risks Report 2020* (2020): 3, fig 2.

**54** Masashi Soga and Kevin J. Gaston, "Shifting Baseline Syndrome: Causes, Consequences, and Implications," *Frontiers in Ecology and the Environment* 16, no. 4 (2018): 222, https://doi.org/10.1002/fee.1794.

**55** Ludovic Slimak et al., "Modern Human Incursion into Neanderthal Territories 54,000 Years Ago at Mandrin, France," *Science Advances* 8, no. 6 (2022):10, https://doi.org/10.1126/sciadv.abj9496.

**56** Elena A. Pearce et al., "Substantial Light Woodland and Open Vegetation Characterized the Temperate Forest Biome before Homo sapiens," *Science Advances* 9, no. 45 (2023): Disc., para. 1, https://doi.org/10.1126/sciadv.adi9135.

## Megafauna Turbocharge Ecosystem Productivity

The megafauna that existed during the Pleistocene helped ancient forests, savannas, and grasslands evolve to support high rates of plant growth by transferring large quantities of limiting nutrients like phosphorous laterally across the landscape from fertile areas such as river flood plains to areas that were less fertile. When many megafauna in Amazonia disappeared over 10,000 years ago, lateral nutrient transfer was estimated to have decreased by 98 percent. This important biogeochemical disruption coincided with the arrival of humans, which prompted some scientists in 2013 to wonder whether the Anthropocene may have begun with human-caused Pleistocene megafaunal extinctions.[57]

The synergistic cooperation of life had created complex ecosystems that were far more biodiverse than those of today.[58] More biodiverse ecosystems tend to be more productive so they would have had larger impacts on local climate. The reality our ancestors experienced was much closer to the Serengeti-like abundance they drew on the cave walls at Chauvet-Pont-d'Arc Cave in France 30,000 years ago than the impoverished landscapes many people live in now (**Figure 4.1**).

Pleistocene megafauna turbocharged the rate of plant growth and cycled essential nutrients like carbon and nitrogen faster. The rate at which photosynthesis happens is important for our future for three reasons. First, more productive ecosystems provide all life (including humans) with more food and resources, i.e., more natural capital. Second, as carbon atoms embedded in plant or algal biomass are consumed by creatures farther up the food chain and get recycled many times, a certain percentage may eventually end up in stable forms of sequestered carbon in soil or ocean sediments, where the carbon can remain sequestered for thousands of years. The higher the productivity, the faster the rate of carbon sequestration. Third, most living things are made of about 40 percent carbon by dry weight. The more biomass that can be supported in an ecosystem, the more carbon it contains. This keeps large quantities of carbon out of the atmosphere.

Life is also composed of about 50 percent water. Fauna are important water reservoirs in dry climates. The large gut capacity of abundant herbivores acts as portable watering tanks for plants, small animals, and insects during the dry season. This water reservoir contained in all plants, animals, fungi, and soil bacteria on green landscapes was an important part of water cycles when life was

**57** Christopher E. Doughty et al., "The Legacy of the Pleistocene Megafauna Extinctions on Nutrient Availability in Amazonia," *Nature Geoscience* 6, no. 9 (2013): 764, https://doi.org/10.1038/ngeo1895.

**58** Steve Webb, "Palaeotrophic Reconstruction and Climatic Forcing of Mega-Lake Eyre in Late Quaternary Central Australia: A Review," *Boreas* 39, no. 2 (2010): 323, https://doi.org/10.1111/j.1502-3885.2009.00120.x.

vastly more abundant in places with semiarid rainfall patterns, helping explain why the planet's hydrology reacted to changes to its biosphere in eons past.

The rain-producing ability of fully functioning ecosystems enabled the past abundance and diversity of life, especially in semiarid areas like the Sahara and Australia. In North Africa, lost species include the bush elephant (*Loxodonta africana pharaoensis*), the North African horse (*Equus algericus*), and the bluebuck (*Hippotragus leucophaeus*), which was hunted to extinction by European settlers around 1800. Three species of giant wildebeest (*Megalotragus spp.*) stood 1.4 meters tall at the shoulder, while their horns spanned 1.2 m tip to tip. North African aurochs (*Bos primigenius mauritanicus*) were giant ancestors of today's cattle that were twice as big. One of the largest herding animals, the African giant buffalo (*Syncerus antiquus*), was widespread in Africa 12,000 years ago and stood almost 2 m tall. Their impressive horns spanned up to 3 m tip to tip, and individuals weighed 1,200 kg on average. Big bulls could have been 2,000 kg, as heavy as a small pickup truck!

Great numbers of these massive North African herbivores, and similar giant grazers across the globe, were the gardeners of forests, grasslands, and savannas. They crowded together for safety in numbers from predators, including the African northern lion (*Panthera leo leo*), Cape lion (*Panthera leo melanochaita*), Barbary leopards (*Panthera pardus pardus*), and Atlas bears (*Ursus arctos crowtheri*). Allan Savory has identified megafaunal behavior as prey animals congregated together for protection from predators as an important contributor in the rapid recycling of nutrients. These animal impacts result in faster plant growth that supports an abundance of life on healthy grasslands and savannas.

## Hourglass-Shaped Food Pyramids

Could the theme of exceptional wildlife abundance depicted in cave art around the world be much more than artistic license? Food pyramids, or trophic pyramids, describe the flow of energy from primary producers, such as plants and algae, at the base of the food chain, up through primary consumers, such as grazing animals. Usually in terrestrial ecosystems, the producer category is represented by a very large base made of up plants, with much smaller trapezoid shapes representing the smaller number and biomass of consumers and predators at the top of the food chain. This is because, as energy flows up the food chain, more energy is used up in the body maintenance of animals and is dissipated as heat, leaving less energy at the top of the food chain. But it is not always so simple.

The cheek-by-jowl depictions of lions and beasts of prey in cave art at Chauvet-Pont-d'Arc in Ardèche, France could reflect elements of an hourglass-shaped trophic pyramid (**Figure 4.1**). These are observed today in ocean ecosystems where large, long-lived consumers and predators, such as whales and sharks, make up a much larger proportion of the food pyramid than expected. On land today, biomass at the base of the food chain builds up in the form of grass and forests, but in pristine ecosystems animals may have cycled this biomass much faster by eating it. The idea that highly abundant wildlife cycled biomass rapidly to create their own abundance was offered by Franz Vera decades ago to explain why tree pollen was more abundant in the fossil record in Europe than grass pollen, even though many fossils of grazing animals could be found. He hypothesized that the herbivores ate the grass and kept it vegetative, so fewer tillers went to seed. This led him to believe open woodland was the pristine ecosystem in most of Europe and not closed canopy forest.[59]

The mind-bending possibility that top-heavy-shaped food pyramids existed on land eons ago becomes more plausible because of an observed law of nature: Kleiber's law. As the average size of a species doubles, members of that species are 25 percent more efficient at maintaining their bodies. This means the metabolic rate per kg of biomass of a 70-tonne whale is only one-tenth that of a mouse.[60] Mammoths needed less food than you'd think.

On pristine "unlimited" savannas that had not yet had their productivity constrained by people, rapidly growing plant food at the bottom of the trophic pyramid fed many more animals in the consumer and predator category than one might expect. This was because large numbers of primary consumers were big herbivores who used food more efficiently. Large-bodied animals tend to live longer as well, so their numbers would have made up disproportionately larger trophic classes of consumers and predators in the food chain compared to today. Because of the unexpectedly high carrying capacity of unlimited ecosystems, the fertilizing effect of "megafaunal impacts," such as dunging and the trampling of old leaves so they could resprout succulent and green, would have been amplified. Lateral transfer of nutrients across the landscape would have turbocharged productivity in the savanna and open woodlands ecosystems around the world as they self-organized into abundance unimaginable by today's status quo.

**59** F. W. M Vera, Grazing Ecology and Forest History (CABI, 2000), 300–307.

**60** Max Kleiber, "Body Size and Metabolic Rate," *Physiological Reviews* 27, no. 4 (1947): 512, https://doi.org/10.1152/physrev.1947.27.4.511.

## The Productivity Paradox of the Mammoth Steppe

The last of several species of mammoth probably lived as recently as 5,000 years ago on Wrangel Island off the coast of Siberia.[61] For millions of years, these woolly relatives of modern elephants roamed large areas of Siberia, Alaska, and Canada. Their territory came as far south as the modern United States as glaciers ebbed and flowed during the cycles of the past eight to 10 glacial events. The superpower that megafauna possessed to increase the productivity of an entire ecosystem is illustrated by the productivity paradox of the mammoth steppe during the last ice age.

How productive was the mammoth steppe before humans limited its productivity? In 2021, a group of scientists led by anthropologist Tylar J. Murchie were researching historical land and sea productivity. They scoured ancient DNA evidence and were surprised to detect an unexpectedly high abundance of animals on the mammoth steppe before humans removed keystone megafaunal species like mammoths and saber-toothed tigers. They uncovered evidence of a cold grassland bustling with so much life compared to the paucity of wildlife the researchers grew up with they found their own conclusions hard to believe.[62]

They cautioned in their paper that the results may not reflect reality and more research is needed. But could their doubts about their scientific conclusions be influenced by their subjective shifting baseline? Did their preconceived beliefs about ecosystem productivity, which came from what they observed on our depleted Earth today, cause them to doubt their own evidence? After all, other scientists found the same surprising abundance. Based on densities of frozen fossil records, Russian scientist Sergey Zimov concluded that mammoth-steppe animal biomass and plant productivity, even in these coldest and driest of the planet's grasslands, were paradoxically high and "the density of animals and their community structure was similar to that of African Savanna." [63]

---

**61** Kh. A. Arslanov et al., "Consensus Dating of Mammoth Remains from Wrangel Island," *Radiocarbon* 40, no. 1 (2006): 294, http://doi.org/10.1017/S0033822200018166.

**62** Tyler J. Murchie et al., "Collapse of the Mammoth-Steppe in Central Yukon as Revealed by Ancient Environmental DNA," *Nature Communications* 12, no. 1 (2021): 7120, https://doi.org/10.1038/s41467-021-27439-6.

**63** S. A. Zimov et al., "Mammoth Steppe: A High-Productivity Phenomenon," *Quaternary Science Reviews* 57 (2012): 41, https:// doi.org/10.1016/j.quascirev.2012.10.005.

## Mammoths Ate Flowers

My pastures at Raindance Farm changed color depending on the month. We often skied on fields with an insulating white winter blanket of snow in January. That gave way to bright green grass in April. By May, yellow polka dots of dandelion flowers erupted, turned white, went to seed in a week, and drifted away in the breeze. Next was the pink splash of red clover and milkweed flowers in June, followed by yellow-orange bird's-foot trefoil flowers. By my father's birthday in August, blue and mustard-colored aster flowers would color the borders of the pastures.

Ancient mammoth steppes may have also grown abundant forbs and changed colors seasonally. Frozen mammoth dung was found to contain the remains of a plethora of flowering plants.[64] Today in springtime, people flock to the savanna grasslands outside San Francisco to see flowers bursting with color on hills once dotted with oaks and browsed by mastodons.

The way supersized herbivores like mammoth crushed tall dominant plants to soil level fostered high levels of plant diversity even on the cold, dry mammoth steppe. Many kinds of colorful flowers bloomed across the region and were readily eaten by mammoth. This could have been the key to the high productivity of the mammoth steppe.[65] Each plant would have had preferred pollinators, other insects, birds, and microbiology associated with it that would have enhanced ecosystem function and help explain the higher-than-expected abundance of life. The more diverse the web of life, the faster the nutrients cycle, resulting in faster-growing plants.

A 2020 study that measured the productivity of perennial plants found that if researchers increased the number of plant types in test plots from one to 60, the standing biomass available to consumers doubled because 50 percent more energy flowed through the ecosystem.[66] The total of these symbiotic and additive relationships is known in ecology as "community dynamics." These learnings can be incorporated into your future vision for a better world since more diversity means more abundance, something you may want your children and theirs to enjoy. You can plant the seeds for a more beautiful and abundant future now by growing more flowers where you live.

**64** Eske Willerslev et al., "Fifty Thousand Years of Arctic Vegetation and Megafaunal Diet," *Nature* 506, no. 7486 (2014): 49, https://doi.org/10.1038/nature12921.

**65** Kari Anne Bråthen et al., "The Paradox of Forbs in Grasslands and the Legacy of the Mammoth Steppe," *Frontiers in Ecology and the Environment* 19, no. 10 (2021): 587, https://doi.org/10.1002/fee.2405.

**66** Oksana Y. Buzhdygan et al., "Biodiversity Increases Multitrophic Energy Use Efficiency, Flow and Storage in Grasslands," *Nature Ecology & Evolution* 4, no. 3 (2020): 394, https://doi.org/10.1038/s41559-020-1123-8.

## The Superpower of Megafaunal Impact

Like wildebeest today, ancient herds of large grazing animals needed to be constantly moving to fresh, unsoiled grass to eat. I learned at Raindance Farm that, surprisingly, the behavior of grazing herds fertilizes more than their dung. Wild grazing herds bunch close together to protect each other from predator attack. Much of the carbon-rich plant material they don't eat, like the tough stems of grasses and woody branches, is trampled into the soil surface by the pressure of this density of animals on the land. The high-carbon plant material complements their high-nitrogen dung, which acts as an inoculant by introducing beneficial bacteria. These ingredients are worked into the soil via hoof action to create a layer of balanced fertilizer in their wake. The plant litter mixed with dung is the ideal compost mix to help hydrate and feed soil biology and grows plants faster (lifts ecosystem productivity) in a positive feedback loop.

Plants contain 38 to 46 percent carbon by dry weight,[67] which is sourced from the air via photosynthesis. By turbocharging nutrient cycling for faster photosynthesis, megafauna helped the land produce more food for themselves and all the other members of the ecosystem. Thus, the grasslands and savannas of the world could have been orders of magnitude more productive than those of today.[68] This means their effect on the climate would have been much stronger.

A 2018 study modeled the productivity of the mammoth steppe during the Last Glacial Maximum. It concluded that size matters: the grazing and trampling by the largest megafauna had an outsized impact on productivity because when considering "the total vegetation as a whole, turnover rate increased by 31 percent, not only because of faster cycling in grass and tree biomass, but also because of the smaller total forest area." [69] Since the last glaciation, wild grazer biomass has been reduced by 79–93 percent due to human hunting of megafauna.[70] This drop in ecosystem productivity would have initiated grassland degradation with far-reaching impacts on the carbon cycle.

67 John S. Pate and David B. Layzell, "Carbon and Nitrogen Partitioning in the Whole Plant—A Thesis Based on Empirical Modeling," in *Nitrogen and Carbon Metabolism* (Springer Netherlands, 1981): 94, https://doi.org/10.1007/978-94-009-8267-3_4.

68 Tobias Andermann et al., "The Past and Future Human Impact on Mammalian Diversity," *Science Advances* 6, no. 36 (2020): 43, https://doi.org/10.1126/sciadv.abb2313.

69 Dan Zhu et al., "The Large Mean Body Size of Mammalian Herbivores Explains the Productivity Paradox During the Last Glacial Maximum," *Nature Ecology & Evolution* 2, no. 4 (2018): 644, https://doi.org/10.1038/s41559-018-0481-y.

70 Zhu et al., "Large Mean Body Size," 641.

From a modern perspective, restoring the lost capacity of grassland and savanna ecosystems to harvest more sunlight is a huge opportunity. The potential to draw down more atmospheric carbon by accelerating plant growth using naturally behaving livestock is huge, especially on desertifying land. Then, if just a small percentage of photosynthetic carbon gets sequestered in deeper soils, the sheer tonnage of carbon that can be sequestered is tremendous! Using modern megafauna, including livestock, as a tool to increase the ability of grasslands to sequester carbon far surpasses any carbon drawdown possible with technology today because it can be applied widely to at least two billion hectares of degraded land. Compared to carbon sequestration technology, little fossil fuel use is required to manage livestock on the land where they belong. The planet-cooling effects of restored ecosystems go far beyond carbon sequestration as water cycles are also restored. Remember, hydrology controls most of the heat dynamics of our planet.

## Shifting Mosaic: A Template for High Productivity

As modern deer hunters know, the ecosystem edges where grassland meets tree cover is a highly productive zone for producing food for wildlife. Large grazers and shrub-eating browsers like woolly rhino, mammoth, and bison once maintained a colorful, highly diverse, shifting mosaic landscape of savanna grasses and flowering forbs. Prior to human arrival, landscapes across Europe were dominated by variable open woodlands of grass, trees, and shrubs.[71] Clumps of oak, hazel, yew, and other light-loving trees peppered the grassy landscape that was kept open by large grazing animals.[72]

The oak savanna became a successful, productive template that Mother Nature applied repeatedly to many bioregions of the Earth. Groves of light-adapted trees like oaks shifted their position on the open grassland, like when the Ents marched in Lord of the Rings: The Two Towers, but in very slow motion, over periods of hundreds of years. In pristine groves, new tree seedlings could not survive the browsing pressure of so many wild animals. Instead, squirrels or birds like blue jays carried acorns and hazelnuts into the grazed meadows nearby and buried them for eating later. If some happened to sprout underneath the protection of a thorny blackthorn or hawthorn bush, they could get big enough to survive grazing pressure and eventually shade out the thorny shrubs and become part of a new grove of mature trees.[73]

**71** Elena A. Pearce et al., "Substantial Light Woodland and Open Vegetation Characterized the Temperate Forest Biome before *Homo Sapiens," Science Advances* 9, no. 45 (2023): Disc., para. 1, https://doi.org/10.1126/sciadv.adi9135.

**72** Elena A. Pearce et al., "Higher Abundance of Disturbance-Favoured Trees and Shrubs in European Temperate Wood-lands Prior to the Late-Quaternary Extinction of Megafauna," Journal of Ecology 112, no. 12 (2024): 2818, https://doi. org/10.1111/1365-2745.14422.

**73** Vera, *Grazing Ecology*, 300–307.

In New Zealand, I witnessed nature's highly productive shifting savanna template in action, but the characters had changed. Here, some Coprosma plants evolved to resist the grazing of the nine species of moa that once roamed in New Zealand. They did this by growing widely angled "divaricating" branches that intertwined in a prickly mess to protect leaves on the inside of the plant. On Guy and Cushla Fraser's regeneratively managed pastoral farm in Taumarunui, a native totara tree in their beautiful woodland has a 3-m diameter. In 2024, I noticed 1-m-tall totara trees sprouting up nearby from under the protective dense branches of Coprosma shrubs in adjacent pastures that had been grazed by both sheep and cattle for decades. Nearby, the fallen remnants of trees that had become old and died were lying in thick grass as the meadow and forest changed places.

## Mimicking Nature at Raindance Farm

How do we build back our rapidly diminishing natural capital in the absence of the giants that helped to co-create it in the first place? Can we mimic the natural fertilizing behavior of wild herding animals that once thronged the plains worldwide with the few megafaunal species left on the planet? Now that livestock make up the vast majority of megafauna, can they be managed much differently to restore naturally functioning landscapes instead of erasing them to grow crops to feed livestock?

Modern megafaunal herbivores on Raindance Farm, our cattle, showed us how changing their behavior to mimic wild grazing herds can result in more plant growth per hectare. Our cows provided "animal impact" by trampling, drooling, dunging, and urinating at various intensities depending on our management. If we let our herd of 85 milking cows and 10 beef steers into a one-hectare field for 24 hours, they would certainly eat half the grass or more and randomly leave dung pats. But this dung would be spread out widely and would not fertilize much. Low-energy weeds that they didn't eat would be left standing to reproduce. I knew that any grass not eaten because a cow had soiled it would be dry and unpalatable by the next grazing event. That would reduce fresh green shoots available for grazing next time and lower pasture productivity.

When we mowed pasture to encourage fresh green shoots to improve quality, the stemmy litter material didn't decompose rapidly enough to become fertilizer, especially in dry summer weather. It sat the width of my hand above the soil surface, held up by uncut grass stems. If the litter did not meet the soil surface, soil biology could not assimilate the plant material for food timely enough and the carbon within it simply started to oxidize into the air in the form of $CO_2$.

But what if we spent 30 minutes splitting up that same one hectare into four smaller paddocks using a single strand of temporary electric fence to feed the cattle in smaller areas at a time? My cows would still get the same 1-hectare area in 24 hours and access the same amount of food as they moved across the landscape, bunched much closer together as their ancestors behaved when predators were all around. Holistic management educator Ian Mitchell-Innes came to our farm in 2011 and taught us that this is how my cows can trample the compost ingredients of uneaten plants and dung down to the soil surface so the earthworms, beetles, and even white tendrils of soil-dwelling fungi could reach into the trampled material and use it for food.

I noticed a completely different aftermath behind the herd when we increased animal density in this way. Cow dung, which was not in sufficient quantity to fertilize much on its own, was scattered more evenly and mixed effectively with the carbon, nitrogen, and minerals contained in the biomass from the trampled plants. The dung acted as an inoculant of microbes for the other compost ingredients to help decomposition. I describe this scenario to my farm clients as "a fertilizer application that must be applied by livestock as close to the soil as a worm can jump." We always followed our grazing events with enough recovery time with no cattle grazing so the plants could regrow. At any one time, most of our land was rested from cattle grazing and available exclusively for wildlife. Only the mild grazing of wild deer, woodchucks, turkeys, and insects occurred at this time.

Our holistic planned grazing management favored more plant diversity in the pasture because livestock were not left in an area for long enough to repeatedly eat the new tender sprouts on the most palatable plants and overgraze them until they die, while letting weeds go to seed. When our desirable grass plants had regrown as many green leaves as they could after a grazing event, we let our cows return to the pasture again. Virtually all the dung and trampled grass had disappeared underground and been used by soil biology. It had been upcycled into a fresh pasture sward with long green leaves. The grass tillers were so thick, sometimes they would trip me up as I walked across the paddocks to put up temporary electric fence subdivisions. In effect, our pasture was a giant solar panel that allowed us to harvest solar energy to convert into milk and then solar dollars to pay our employees and the mortgage. In our case, holistic planned grazing helped lift our farm's profitability because it helped us avoid either overgrazing or over-rest of pasture plants, both of which would have resulted in lower pasture productivity.

## Farms Sequester Carbon

Ecosystem integrity and productivity enhancement by large herbivores is central to solving our modern predicament of having much more $CO_2$ in the air than we have had for the last 800,000 years.[74, 75, 76, 77, 78, 79, 80] The productivity-boosting effect of increased plant and animal diversity can help humans design more-productive agroecosystems. A recent review of 22 scientific papers on the effect of combining livestock and temperate agroforestry systems showed increased productivity per hectare. When sheep were paired with tree crops (silvopasture), 1 hectare of silvopasture produced the same number of livestock units and silvicultural products as 0.74 hectares of pasture plus 0.70 hectares of forest grown separately.[81] In other words, one hectare of silvopasture had the same productivity as 1.44 hectares of each grown alone. The study even detected an additional boost to productivity when cows were included instead of sheep, a nod to the effectiveness of megafaunal impact as grazing animals get larger.

---

**74** Yu Feng et al., "Doubling of Annual Forest Carbon Loss over the Tropics During the Early Twenty-First Century," *Nature Sustainability* 5, no. 5 (2022): 448, https://doi.org/10.1038/s41893-022-00854-3.

**75** Jeppe A. Kristensen et al., "Can Large Herbivores Enhance Ecosystem Carbon Persistence?," *Trends in Ecology & Evolution* 37, no. 2 (2022): 125, https://doi.org/10.1016/j.tree.2021.09.006.

**76** Dilip G. T. Naidu et al., "Loss of Grazing by Large Mammalian Herbivores Can Destabilize the Soil Carbon Pool," *Proceedings of the National Academy of Sciences of the United States of America* 119, no. 43 (2022): Disc., para. 1, https://doi. org/10.1073/pnas.2211317119.

**77** Silvio Daniele Oggioni et al., "Livestock Grazing Abandonment Reduces Soil Microbial Activity and Carbon Storage in a Mediterranean Dehesa," *Applied Soil Ecology* 153 (2020): sect. 5, 103588, https://doi.org/10.1016/j.apsoil.2020.103588.

**78** E. F. Viglizzo et al., "Reassessing the Role of Grazing Lands in Carbon-Balance Estimations: Meta-Analysis and Review," *Science of The Total Environment* 661 (2019): 538–540, https://doi.org/10.1016/j.scitotenv.2019.01.130.

**79** Rattan Lal et al., "The Carbon Sequestration Potential of Terrestrial Ecosystems," *Journal of Soil and Water Conservation* 73, no. 6 (2018): 148A–149A, https://doi.org/10.2489/jswc.73.6.145A.

**80** Yi Yang et al., "Soil Carbon Sequestration Accelerated by Restoration of Grassland Biodiversity," *Nature Communications* 10, no. 1 (2019): Disc., para. 3, https://doi.org/10.1038/s41467-019-08636-w.

**81** Gabriel J. Pent, "Over-Yielding in Temperate Silvopastures: A Meta-Analysis," *Agroforestry Systems* 94, no. 5 (2020): 1751, https://doi.org/10.1007/s10457-020-00494-6.

Our farm soil tests showed increased organic matter measurements, and this taught us that when plants grow faster, they secrete more liquid carbon deep into the soil profile, where it is more likely to remain as stable carbon, increasing our opportunity to sequester carbon.[82, 83] Other farmers have seen their soil carbon increase with holistic planned grazing, and it has been recognized in the literature as one way to help mitigate climate change.[84, 85]

## Embarking on the Restoracene

You can begin to repair legacy damage and rebuild natural capital with holistic decisions around what you eat and what you buy. Along with conservation of resources, including fossil fuels, this can ensure your grandchildren will have the water, food, oxygen, energy, and shelter they need. All the tools available to people for managing nature require creativity to be implemented, so it is unlikely artificial intelligence will usurp the need for people to manage the resources that make up our natural capital. As Allan Savory and Jody Butterfield write in *Holistic Management*, "Every situation requires management that must be an original product of human imagination, and even that must evolve as the situation changes."[86] But everyone needs to understand ecology to make sensible decisions.

---

**82** Kristensen et al., "Can Large Herbivores," 125.

**83** Noah W. Sokol et al., "Evidence for the Primacy of Living Root Inputs, not Root or Shoot Litter, in Forming Soil Organic Carbon," *New Phytologist* 221, no. 1 (2019): 242, https://doi.org/10.1111/nph.15361.

**84** Adam D. Sacks et al., "Reestablishing the Evolutionary Grassland-Grazer Relationship to Restore Atmospheric Carbon Dioxide to Preindustrial Levels," in *Geotherapy*, eds. Thomas J. Goreau, Ronal W. Larson, and Joanna Campe (CRC Press, 2014), 167–178, https://doi.org/10.1201/b13788-9.

**85** W. R. Teague et al., "The Role of Ruminants in Reducing Agriculture's Carbon Footprint in North America," *Journal of Soil and Water Conservation* 71, no. 2 (2016): 159–162, https://doi.org/10.2489/jswc.71.2.156.

**86** Allan Savory and Jody Butterfield, *Holistic Management: A Commonsense Revolution to Restore Our Environment* (Island Press, 1999): 169.

# Chapter 5

# The Rise of the Superpredator

I always loved my evening pasture walks on Raindance Farm to check on the cows. My dog Lucero usually trotted ahead of me unless it was calving season when I didn't bring her so as not to upset the new moms. Barn swallows performed acrobatics over our heads and deer emerged from the forest to graze in our pastures. As evening dew fell, the cows would start to lie down after filling their bellies with sweet grasses, legumes, and forbs in preparation for a night's work of ruminating. This diligent rechewing of regurgitated cud (partially digested grass) transforms tough grass leaves into simpler compounds via microbial fermentation in the cow's stomach. When the rechewed cud is swallowed again, it becomes food for the microbiome residing in her rumen, the largest of her four stomachs. These break it down into usable energy and fatty acids, the building blocks for meat and milk.

One evening out with the cows, a surprise encounter got me thinking about relationships between dogs and humans. I was startled by a deer that suddenly jumped up from the pasture only three meters away from us and bounced away into the nearby forest cover. She had been nestled down in knee-high grass hoping we would walk by and not see her. Some wisdom of nature was revealed to me because my dog was loyally looking up at me, tongue hanging down, with absolutely no idea of what had happened. The grass was too tall for her to see the deer, and she had not smelled anything because of the direction of the wind. It made me wonder if humans developed an evolutionary advantage by standing up on our hind legs because we could see game and ever-present predators better in the tall grass of the African savannas where we evolved.

Humans lack the keen sense of smell or hearing that other animals possess to easily detect game. Did dogs develop such a keen sense of smell and hearing because they couldn't see far in tall foliage? My farm had become my teacher and started me down an investigative path, where I discovered our species had teamed up with dogs starting some 30,000 years ago. As humans coevolved with dogs at our side, together we became superpredators.

Realizing that our human ancestors had the ability to cause megafaunal extinctions will help humanity own the magnitude of our role in widespread ecosystem decline. Because the rate of species extinction is ramping up today, understanding how it occurs is the only way we can reverse this age-old trend. Again, the root of the problem is our Stone Age tendency to make shortsighted decisions that solve the problem at hand, such as what is for dinner, but can result in catastrophic unintended consequences.

The many species of animals our ancestors stole from our generation, such as woolly mammoths, mastodons, and giant sloths, had survived many previous glaciations over 2.6 million years. Some of the climatic changes during previous glaciations were even more abrupt than those associated with the last cold period. It makes no sense that so many large species could not survive these same changes 15,000 years ago, so climate change is unlikely to be the cause for megafaunal extinctions.

Scientists have observed from spore and pollen data that the abundance of large herbivores in many locations of the US declined prior to changes in vegetation. These changes were so significant they even increased the occurrence of wildfire. This could mean that when grazing pressure from megafauna became less, more trees were able to grow over vast areas. These landscape changes were not driven just by climate as once widely believed.[87]

Alternatively, human-caused extinctions may have contributed to the decline of ecosystems worldwide that eventually impacted climate, and not the other way around. There is a correlation between the arrival of Homo sapiens and unnatural rises in extinction rates all around the globe, particularly the timing of human arrival to Australia 65,000 to 44,000 years ago and to the Americas 24,000 to 12,000 years ago.[88]

---

**87** Angelina G. Perrotti et al., "Diverse Responses of Vegetation and Fire After Pleistocene Megaherbivore Extinction Across the Eastern US," *Quaternary Science Reviews* 294, (2022): sect. 5, 107696, https://doi.org/10.1016/j.quascirev.2022.107696.

**88** Tobias Andermann et al., "The Past and Future Human Impact on Mammalian Diversity," *Science Advances* 6, no. 36 (2020): Causes of extinctions, para. 2, https://doi.org/10.1126/sciadv.abb2313.

## Human Impact on the Mammoth Steppe

Why did our ancestors hunt the mammoth steppe in Siberia, Alaska, and Canada's Yukon territory in the first place? Surely, it was a cold, inhospitable place most of the time? Couldn't they have migrated to warmer subtropical climates, especially during the last glacial period, and eaten fruit, nuts, and tubers? It turns out the economic reward for effort put into hunting mammoths and giant bison was very high.

Large herbivores, like mastodons of old and today's cattle, can turn high-carbon-content plant material that is inedible for humans into highly nutritious fat that has twice the energy of protein or carbohydrates per gram. One mammoth kill could feed a whole tribe of Paleolithic people for a considerable period. Could nomadic summertime hunting excursions to the mammoth steppe have allowed humans to put on fat reserves to help them withstand the long ice age winters? Could the abundance of high-energy food have wooed people to stay longer?

Large grazing animals evolved to accumulate body fat that would help them survive during periods of food scarcity, such as winters in cold climates or the dry seasons in semiarid climates. The high energy density of megafauna meat, combined with the amount of meat harvested per hour of hunting, may have attracted the early humans to hunt the largest animals. Examination of the human remains of an 18-month-old Clovis boy[89] in what is now Montana revealed that the diet of his mother 13,000 years ago was 40 percent mammoth meat, with another 10 percent made up of other megafaunal species such as bison, elk, horse, and camel.[90]

Paleolithic people hunted the largest game available. For example, humans have lived in the Levant region for at least 40,000 years. A study of the remains humans left behind revealed that people hunted the largest species first, and when they were depleted, worked their way down the body size distribution over tens of thousands of years.[91] This appears to be the pattern all over the globe as humans settled new lands.

---

89 The Clovis people were a culture in North America known for their stone-tipped weapons. They left evidence that is 13,050–12,750 years old.

90 James C. Chatters et al., "Mammoth Featured Heavily in Western Clovis Diet," *Science Advances* 10, no. 49 (2024): Disc., paras. 1–2, https://doi.org/10.1126/sciadv.adr3814.

91 Jacob Dembitzer et al., "Levantine Overkill: 1.5 Million Years of Hunting Down the Body Size Distribution," *Quaternary Science Reviews* 276, 107316 (2022): 7, https://doi.org/10.1016/j.quascirev.2021.107316.

## How Did Paleolithic People Wipe Out Megafauna?

When the Australian megafaunal extinctions took place, around 40,000 to 50,000 years ago, all except two species weighing over 45 kg disappeared. Gone forever are five reptilian mega-predators reminiscent of today's Komodo dragon, a marsupial lion, and the world's largest wombats and kangaroos. A scientific team led by Sander van der Kaars implicated humans as the drivers of megafaunal extinction in Australia in a study that ruled out climate factors. This paper goes further, offering insights on the plausibility of megafaunal collapse due to human hunting and the theory of "imperceptible overkill."[92]

The imperceptible overkill theory could help explain human-orchestrated megafaunal extinctions globally.[93] Large mammals are vulnerable to hunting pressure because they typically reproduce slowly. It has been estimated that low-intensity hunting (such as the killing of just one juvenile per person per decade) could result in a megafaunal species being extinguished over a few hundred years. The examination of fossils at Cape Pasley, Australia, revealed that megafaunal population collapse started within 2,000 years of the established date for human arrival to the southern part of Western Australia about 47,000 years ago. Extinction was completed within 4,000 years, just a blink in evolutionary time.[94]

It did not take many people to cause so many large species to disappear because humans have been a highly effective predator for millennia. Nowhere was this more pronounced than in New Zealand, the only place in the world where megafaunal extinction was 100 percent complete. Around eight hundred years ago, all nine species of the giant feathered but flightless moa were eliminated from New Zealand within just 120–150 years of human arrival. Scientists think it only took between 1,500 and 2,500 people. This occurred using primitive but effective weapons against giant birds that towered above the height of their hunters.[95]

**92** Sander van der Kaars et al., "Humans Rather than Climate the Primary Cause of Pleistocene Megafaunal Extinction in Australia," *Nature Communications* 8 (2017): Disc., para. 4, 14142, https://doi.org/10.1038/ncomms14142.

**93** Yadvinder Malhi et al., "Megafauna and Ecosystem Function from the Pleistocene to the Anthropocene," *Proceedings of the National Academy of Sciences* 113, no. 4 (2016): 839, https://doi.org/10.1073/pnas.1502540113.016

**94** Van der Kaars et al., "Humans Rather than Climate," Disc., paras. 1–2.

**95** Richard N. Holdaway et al., "An Extremely Low-Density Human Population Exterminated New Zealand Moa," *Nature Communications* 5, no. 1, 5436 (2014): 5–7, https://doi.org/10.1038/ncomms6436.

## A Lethal Weapon Game Didn't Run Away From

Stone Age peoples developed hunting technologies that could be launched to kill from a distance, like arrows and spears. Some spears were equipped with an atlatl, also known as a woomera in Australia, a wooden rod with a hooked end that effectively added another joint and limb to the human throwing arm. This created more thrusting power to fatally injure the largest game without getting close enough to make the game run away or possibly risk having the hunters get hurt. The Guinness World Records distance for an atlatl throw is over 848 ft (258 meters), or greater than the length of two football fields. A projectile spear point made from bone was found embedded in a mastodon rib in Washington State that is 13,800 years old.[96]

Tens of thousands of years ago, abundant game animals were easier to hunt than you might think on continents outside of Africa. As our ancestors left Africa to explore new worlds, they encountered faunal diversity and abundance that was beyond our wildest dreams. These wild, woolly, monster-sized herbivorous animals were probably reluctant to run away from human predators until they got very close, and this could have been their downfall.

The "flight zone" is the space around prey animals that they cannot tolerate any predator entering. Many farmers have learned to use it to their advantage in moving cattle in less stressful ways, as developed by the late Bud Williams, a guru of low-stress livestock handling. He taught a calm approach to moving cattle using mindful placement of the herder around the herd, and certainly not the loud shouting associated with the cowboys working with herds in Western movies.

If a predator gets too close and enters the animal's flight zone, prey animals move away. But they don't run as soon as they see a predator in the distance, because in the pristine ecosystems of the past that teemed with life, predators were abundant and a common sight. To conserve energy, grazing herds evolved to run away only when a predator gets close enough to attack. When humans arrived on the scene, large megafauna probably watched us approaching them but didn't understand we could kill them at a much greater distance than other predators because of our projectile weapons. This could explain why individual animals in surviving deer species take flight as soon as they see, hear, or smell us, as this behavior prevented their species from going extinct like the giant Irish elk and 11 American antelope species.

---

**96** Michael R. Waters et al., "Pre-Clovis Mastodon Hunting 13,800 Years Ago at the Manis Site, Washington," *Science* 334, no. 6054 (2011): 351, https://doi.org/10.1126/science.1207663.

## Colluding with Canine Killers

Social cooperation and technology around hunting helped make humans the dominant megafauna predator around the globe, but it took more than technology to become a superpredator. It is feasible that people could have hunted slow-growing and slow-moving giant herbivores to extinction, but what about horses? Surely the now-extinct American Paleolithic horses were fast enough to flee from danger. Were they so curious about the new human predators that they came within range of accurately thrown atlatl spears? Wouldn't they learn to run after they witnessed their relatives get killed, or did humans have another ace up their sleeve? Several species of early horses, including *Equus ferus caballus*, roamed the Americas for millions of years before humans arrived, but they disappeared from what is now Alaska and Canada's Yukon territory soon after people came and had devoured most of the mammoth. Human coevolution with dogs may have assisted humans in hunting the agile and swift Paleolithic horses, hastening their extinction.

Dogs were domesticated in Siberia 23,000 years ago, and they traveled with the first explorers over the land bridge to the Americas at the end of the last ice age. They quickly dispersed throughout North and South America with humans 15,000 years ago.[97] Evidence of seven horses butchered by humans 13,300 years ago was found at Wally's Beach, Alberta, Canada. They may have been cornered and killed—a tactic long used by Paleolithic humans to hunt horses in Eurasia. Many other archeological sites in Alberta link horses and human hunters at that time. American horses disappeared from the fossil record in the Americas by 10,000 years ago, although DNA remnants found in sediments hint that some may have persisted in remote regions of the Yukon territory in Canada as late as 5,000 years ago.[98] All horses that exist in the Americas now are a different species introduced from Europe.

My partner, Rick, is a shepherd at my new home in New Zealand, where he and his son take care of 4,200 sheep on a 440-hectare farm. We have adopted a border collie from a family in town who wanted her to have more space to run off her pent-up energy. Even as an adult dog who was never trained to herd livestock, her natural drive to find sheep and bring them to me as I waited by the gate into a fresh pasture was obvious right away. She was easy to train to herd sheep, moving large flocks of ewes and lambs by herself to new pastures.

**97** Angela R. Perri et al., "Dog Domestication and the Dual Dispersal of People and Dogs into the Americas," *Proceedings of the National Academy of Sciences of the United States of America* 118, no. 6 (2021): Conclusion, para.1, https://doi. org/10.1073/pnas.2010083118.

**98** William T. Taylor, *Hoof Beats: How Horses Shaped Human History* (University of California Press, 2024), 26.

Teamed up with the other working dogs on the farm, they can shift large mobs of sheep efficiently and love to do it so much that we often need to hold them back!

The strong instinct of modern working dog breeds to bring animals toward their human masters may have started with their ancestors. With a little training, dogs are naturally driven to bring animals or livestock toward the handler, like wolves bring game animals toward the alpha animal in the pack. The arrangement was probably mutually beneficial for both dogs and humans. If dogs steered game animals close to humans during our coevolution, they lowered their risk of being injured or killed by deadly flying hooves or horns, as the responsibility for the kill itself would fall to the humans. Paleolithic people would dispatch the game from a safe distance with the help of their hunting technology, and the dogs would be rewarded with some of the spoils from a successful hunt to reinforce this team effort. Your dog may beg at dinnertime, just like his ancestors did tens of thousands of years ago as they waited for their share of the hunt.

Our human eyesight performs well at a distance, having been honed by millennia spent hunting on the open savanna. In comparison, our young collie's eyesight is poor at a distance. She often cannot see where the sheep are from her low vantage point in our rolling hill country. She can, however, understand commands by voice or whistle to run in any direction we tell her, with the job of retrieving the right group of sheep, even if she can't see the flock we want her to bring to us until she gets closer to it.

When human and dog predators work cooperatively, the result is a superpredator that may have been the demise of swift American horses, along with countless other species. This relationship combines exceptional human eyesight and the unfair advantage of projectile weapon technology with the great sense of smell, speed, and teeth of dogs. When I suggest this theory to my ranching friends who have livestock dogs, this symbiosis makes perfect sense to them because they work together cooperatively with their dogs every day. They understand the innate eagerness of their working or hunting dogs to team up with them. But is there any evidence of our ancient ancestors hunting with the help of dogs?

Rock art found in the deserts of Saudi Arabia depicts hunters aiming bows and arrows at wild asses (*Equus africanus*). In **Figure 5.1**, sixteen dogs are shown surrounding an equid, a common theme of rock art of the Shuwaymis region in Ha'il Province, Saudi Arabia, that has been dated to be around 8,000 years old.[99] Perhaps the game were herded closer to humans by cooperative dog packs?

**99** Maria Guagnin et al., Pre-Neolithic Evidence for Dog-Assisted Hunting Strategies in Arabia, *Journal of Anthropological Archaeology*, 49 (2018): 233, fig. 9, https://doi.org/10.1016/j.jaa.2017.10.003.

**Figure 5.1 Eight-thousand-year-old rock art at Shuwaymis, Saudi Arabia, shows a bow hunter accompanied by dogs with equid game facing the hunter.**

Image by Huw Groucutt as part of the Paleodeserts Project (2015). Source: Guagnin et al., 2018.

## Hunting for Fun

Humans had ample justification to kill for their food to feed their families. Hunting pressure for basic survival may have changed entire landscapes because removing keystone species—especially large ecosystem engineers—caused a domino effect leading to ecosystem decline. But there is also evidence of "wasteful hunting" or hunting just for fun, practiced by our ancestors.

As wildlife became increasingly scarce due to overhunting around ancient Rome, staged hunts rather than actual hunts were put on display by Roman elites to impress slaves and the working class. William Epplett's 2001 PhD thesis, *Animal Spectacula of the Roman Empire*, explores detailed ancient records documenting how wild animals from far-flung regions of the Roman Empire were slaughtered for pure entertainment during *venationes bestiarum*, or staged beast hunts. These were enacted in conjunction with gladiator fights or on their own, sometimes taking 10 days to complete the slaughter. The pompous events put on by politicians were paid for by taxing the poorer citizens who the displays were meant to impress.

Thousands of large wild mammals, including Asiatic lions, North African bears, and elephants, were captured in North Africa and Syria by Greeks and Romans for violent public events. The callous attitude of humans towards wildlife is epitomized by the discovery of the bones of 50 large lions, panthers, leopards, and tigers that were excavated from the sewers beneath the Colosseum in Rome and the surrounding area.

A network of trade in exotic wildlife meant elites throughout the Roman Empire, as far away as France and Switzerland, could buy wildlife to fight other animals or be killed by gladiators in amphitheaters.[100] At an event organized by Pompey in 55 BC, 20 elephants, 500–600 lions, 410 leopards, apes, and a rhinoceros were slaughtered. In 25 BC, a Roman magistrate hosted a venation where 300 bears and 300 other African wildlife were killed for entertainment. Augustus, the first Roman emperor, boasted of having killed 3,500 animals during the 26 venationes bestiarum Africanorum of his reign, including 460 lions.[101] The horrific slaughter of scores of confined animals for spectator sport and political favor continued in many regions of the so-called civilized world until Emperor Anastasius I banned beast hunts in 498 AD.[102]

Other nobility and politicians in India, Egypt, and the Middle East also enjoyed the outrageous slaughter of wildlife for sport. The Egyptian pharaoh Ptolemy II Philadelphus (r. 284–246 BC) founded the cities of Berenice Troglodytica and Ptolemais Theron on the Red Sea coast of Africa as hunting bases. He captured African elephants for use in his military campaigns even though they were not suited to being trained as such. Over-hunting depleted elephant populations along the Red Sea coast and northern Somalia in less than three decades. These semiarid landscapes, impoverished of their greatest wild fauna, would have begun slipping into ecosystem decline and desertification as a result.

## The Grace of Pure Nature

Humans continue to traffic wildlife and destroy non-human life at an ever-accelerating pace, wrecking the natural beauty the Creator gifted to us. Once society realizes that humans are part of nature, we will understand that the destruction of nature is the destruction of ourselves. How civilization has changed, and yet many young people have lost their once respected places in society, many even resorting to gang affiliation or suicide. Could meaningful purpose help ground the hordes of young people involved in gangs and conflicts today? What if they reconnected with nature. What if they had opportunity to be educated in ecology and holistic management and rewarded financially for restoring their homelands? Wouldn't everyone be better off?

---

**100** Jordon Alex Houston, "Exotics for Entertainment : A Reconstruction of the Roman Exotic Beast Trade ( First to Third Centuries AD )," *Theoretical Roman Archaeology Journal* 7, no. 1 (2024): 11, doi: https://doi.org/10.16995/traj.10158.

**101** William Christopher Epplett, "*Animal Spectacula of the Roman Empire*" (PhD diss., University of British Columbia, 2001): 31–32.

**102** Epplett, "*Animal Spectacula,*" 37.

Equipped with our modern-day hubris, humans like to think we have continued to evolve for the better since the days of our hunter-gatherer ancestors. But isn't abundant good food, clean water, security, meaningful social interaction, and a good bit of fun what we all want and need to be happy? Were the unlimited grasslands of our ancient past the proverbial Garden of Eden, albeit more exciting, dangerous, and unpredictable, and with real monsters to contend with? Many people feel something is missing in modern life, and billions of us are searching for something meaningful. Addressing the state of the environment, the late Pope Francis remarked in his 2015 encyclical *Laudato Si': On Care for Our Common Home:*

> A sober look at our world shows that the degree of human intervention, often in the service of business interests and consumerism, is actually making our earth less rich and beautiful, ever more limited and grey, even as technological advances and consumer goods continue to abound limitlessly. We seem to think that we can substitute an irreplaceable and irretrievable beauty with something which we have created ourselves.[103]

You are a predator. Like the tiger, you are the grassland and the forest. As *Homo sapiens*, we are part of the evolutionary trajectory of our planet's biosphere, inseparable from all other species. But, living in our modern caves of treasures, we have removed our bodies and souls from the natural world that spawned us. This is why farmers, hunters, and outdoor enthusiasts love to be on the land. It is where we belong. It is in our soul.

No, I am not saying we should revert to the Stone Age! The ecosystems around us are forever changed, so there is no going back, only forward. But we can choose to move toward a future we want, not the future dictated by the marketers of products that are filling up landfills and polluting our air, land, and water. Living in harmony with the whole of humanity, who are made up of highly educated and cultured people living in comfortable, efficient shelters optimized with the latest technology, would be a good option. Who knows what wonders human creativity can come up with for efficient and ecological cities, economies, and adventures if we try. Your utopia may be different from mine, but if we all are working in good faith to take care of our Mother Earth and each other, Homo sapiens can thrive for another two million years thanks to us. We will only succeed if we take responsibility for passing on the goodness of life to future generations. We can enable evolution to resume so life can adapt as is required by the geoclimatic conditions going forward.

---

**103** Francis, *Laudato Si': On Care for Our Common Home* (Vatican Press, 2015), para. 34, https://www.vatican.va/content/francesco/en/encyclicals/documents/papa-francesco_20150524_enciclica-laudato-si.html.

The time has come for humanity to move into the Restoracene. The start of this epoch marks the first time in human history our species uses tools to make decisions that both feed us and result in better environmental and social outcomes. The Anthropocene is not new; it has been going on for 64,000 years, but it must end now, or our children, and theirs, will suffer terribly. Sadly, we could destroy many other irreplaceable earthling species with us if we continue to manage in the reductionist way our cave-dwelling ancestors did and refuse to evolve. As we manage natural resources, societies, and economies, we must multitask and consider the quality of life we really want for our children and grandchildren.

# Chapter 6

# The Biosphere is Linked to the Atmosphere

Considering the deplorable 70 percent global decline in wildlife over the past 30 years,[104] it behooves us to understand the outcomes of previous human-caused species extinctions. Ancient pyramids and abandoned stone cities worldwide stand testament to collapsed civilizations where intensive agriculture led to soil erosion and reduced agricultural output.[105] Long before farming, however, humans were causing damage to entire ecosystems by selectively hunting giant keystone species.

The research results I have unearthed from often separate scientific disciplines are like individual points of light in a dark sky. By connecting them, a picture of our past appears, just like a constellation in the night sky can outline a recognizable figure when individual stars are connected in the mind's eye. The outline that emerges when archaeology, paleoclimatology, paleoanthropology, ecology, zoology, microbiology, and geology are integrated is the sheer magnitude of human agency to impact global climate by how we manage nature.

In the past, humans have wielded our dominance in many ways that damage the biosphere that we are embedded in. But my experiences at Raindance Farm taught me that we can achieve the opposite outcome, even as we produce resources like food, fiber, and lumber. Not until I saw ground-nesting bird numbers explode on our farm did I realize that farmers (or anyone that eats) could work toward healing the planet, starting with their next holistic action or consumer decision. We all know we are damaging our life-giving biosphere, and most people will not change their maladaptive behavior until they change how they view themselves as powerful regulators of climate, with or without fossil fuels.

**104** WWF (World Wildlife Fund), *Living Planet Report 2022*: Building a Nature-Positive Society (2022).

**105** Marco. A. Janssen and Marten Scheffer, "Overexploitation of Renewable Resources by Ancient Societies and the Role of Sunk-Cost Effects," *Ecology and Society* 9, no. 6 (2004): 1–3.

## The Anthropocene

The Anthropocene is the geological epoch during which human activities have significantly impacted the planet's global systems. From the flow of rivers to the climate, human impact has become evident in geological records. Recent research suggests that the Anthropocene may have started well before 1952, the date proposed by the Anthropocene Working Group, because of the negative effects humans have had on climate stabilizing ecosystems by removing millions of animals.[106, 107]

One of the first scientists to suggest an early Anthropocene was William Ruddiman, a soil scientist who proposed that humans have influenced the global climate for thousands of years through the unintended consequences of intensive farming. In a 2003 paper, he suggested that greenhouse gas emissions from agriculture and land-use changes have prevented the Earth from heading toward the next glaciation, causing the climate to diverge from the patterns seen over the past 2.4 million years.[108]

The orbital cycles that have dictated the strength of the sun's energy in the northern hemisphere over the past two million years have caused the ebb and flow of many glacial cycles during the ice age. The sun's heating effect has a greater impact in the northern hemisphere than in the southern hemisphere because large glaciers over Canada and Greenland act like an icebox for the global climate. The combination of the Earth's distance from the Sun, its tilt on its axis that faces the northern hemisphere toward or away from the sun, and the shape of the Earth's orbit have been collectively named the Milankovitch cycles. The phenomenon is named after the scientist who figured out how the alignment of the Sun and the Earth affects climate patterns, including the onset of cold glacial periods and the warm interglacial periods between them.

The sun's intensity upon the ice fields of Greenland and Canada started to decrease 8,000 years ago, which should have expanded glaciers and cooled the climate. After all, this is what happened during the last eight or ten glaciations of the past two million years, a period known as the ice age. Instead, according to Ruddiman, an abrupt global warming trend around 8,000 years ago occurred due to the release into the air of $CO_2$ and methane from disturbed soil and burned

---

**106** Christopher E. Doughty et al., "The Legacy of the Pleistocene Megafauna Extinctions on Nutrient Availability in Amazonia," *Nature Geoscience* 6, no. 9 (2013): 764, https://doi.org/10.1038/ngeo1895.

**107** David K. Wright, "Humans as Agents in the Termination of the African Humid Period," *Frontiers in Earth Science* 5 (2017): 121, https://doi.org/10.3389/feart.2017.00004.

**108** William F. Ruddiman, "The Anthropogenic Greenhouse Era Began Thousands of Years Ago," *Climatic Change* 61, no. 3 (2003): 288, https://doi.org/10.1023/B:CLIM.0000004577.17928.fa.

forests, as rice and grain cultivation expanded in the Middle East and Asia.[109] As we discussed earlier, ice core evidence from Antarctica shows that atmospheric $CO_2$ and methane are positively correlated with global temperature.[110] Ruddiman suggests that carbon lost from degrading soils oxidized into $CO_2$ that wafted into the atmosphere to mitigate any temperature drops from the decreasing insolation, and helped keep our planet's temperature steady for thousands of years.

## Animals Co-create Grassland Ecosystems

The natural behavior of grazing and plant trampling by herbivores is essential to prevent grasslands in semiarid parts of the world from turning into desert, a process called desertification. This insight was first revealed by Allan Savory in his books[111, 112, 113] and his 2013 TED talk.[114] In seasonally dry savanna and grassland, bunch-forming perennial grass plants produce a mass of leaves during the rainy season. During the dry season, the uneaten leaves turn light brown. If no animals eat or trample these leaves before the next rain, they block sunlight from reaching the growing points at the base of the plant. If these dead leaves block light long enough and prevent photosynthesis, the plant will die even if it rains.

The death of individual plants is the first step in desertification because it eventually reveals bare soil. Without the protection of wind-blocking leaves and soil-binding roots, the soil is vulnerable to wind and water erosion, and it is washed or blown away. Bare soil heats up to far higher temperatures than soil shaded by plants. By reducing the abundance of grazing herbivores on ancient Australian plains, the first peoples there unintentionally started the process of desertification just like their human relatives were doing in Southeast Asia and the Arabian Peninsula.

The pastoralists in the Sahara who drew cave art portraits of themselves and their cattle 4,000 years ago may have contributed to its desertification (**Figure 6.1**). So, how grazing happens is strategically important to the health of grassland ecosystems. In the absence of predators, which occurs when livestock are protected by people, grazing animals tend to both overgraze some plants and over-rest other plants in an area; both processes lead to plant death and subsequent desertification.

---

**109** Ruddiman, "The Anthropogenic Greenhouse Era," 274.

**110** Petit et al., "Climate and Atmospheric History," 431, fig. 3.

**111** Allan Savory and Jody Butterfield, *Holistic Management: A Commonsense Revolution to Restore Our Environment*, 3d ed. (Island Press, 2016): 34–49.

**112** Jody Butterfield et al., *Holistic Management Handbook: Regenerating Your Land and Growing Your Profits*, (Island Press, 2019): 130–133, https://doi.org/10.5822/978-1-61091-977-7.

**113** Allan Savory, *The Grazing Revolution: A Radical Plan to Save the Earth*, 2d ed. (Savory Institute, 2023).

**114** Allan Savory, "How to Fight Desertification and Reverse Climate Change" TED Talk, Long Beach, CA, USA, March 4, 2013, video, 8:34–12:40, https://www.ted.com/talks/allan_savory_how_to_fight_desertification_and_reverse_climate_change.

**Figure 6.1 4,000-year-old pastoral scene in Africa.**

Reprinted from Coulson and Campbell, 2010.

The tool of holistic planned grazing on my farm, and thousands of others, has been effective in reducing both overgrazing and over-resting of pasture plants. Many ranchers in arid locations have used holistic planned grazing to stop desertification in its tracks and induce a resurgence of the diverse savanna that once thrived. A prime example can be seen today at Dimbangombe, the ranch in Zimbabwe given by Allan Savory to the local community to be run as a learning site: the Africa Centre for Holistic Management. The landscape at Dimbangombe is grazed infrequently by cattle and goats at high animal densities by local herders. The result is a much greener and wetter landscape than even the national parkland nearby that is desertifying rapidly. Streams run more consistently, and wet holes not seen in modern lifetimes are making a comeback. This supports more wildlife with fodder and water.

Across the Atlantic in Mexico, Las Damas Ranch is holistically managed by Alejandro Carrillo. On his family's approximate 10,000 grazable hectares in the Chihuahuan Desert, they are growing thicker and taller grass, whereas before they began using holistic management, bare soil and mesquite bush was the norm.[115]

115 "Las Damas Ranch Case Study," Understanding Ag, accessed July 25, 2025, https://understandingag.com/case_studies/las-damas-ranch-case-study/.

Carrillo told me in a 2025 interview that the ranch can run five times as many livestock as the regional average stocking rate and still have plenty of grazing for wildlife.[116] Their cattle normally only graze each paddock once every 18 months, which is producing amazing results in such a dry climate. He says on farm monitoring had seen diversity go from 20 to 100 different kinds of grassland plants.[117] These are just a few examples of how grazing can be managed to enhance ecosystem productivity instead of degrading it.

## Humans Initiate the Decline of the Sahara

It is hard to imagine the vast blowing sands of the Eastern Saharan landscape as anything but desert. But in 1933, the Austrian explorer László Almásy ventured deep into the parched Gilf Kebir–Jebel Uweinat region and discovered a cave with rock art depicting ancient people swimming! At the Cave of Swimmers, made famous in a scene from the film *The English Patient*, drawings across the full length of the rock shelter show tall adults with physiques like the Na'vi from the *Avatar* films seemingly watching over children swimming. The cave was populated by hunter-gatherer people as long as 7,500 years ago, but by 6,400 years ago cattle started to appear in the rock art. By 5,500 years ago this region where Libya, Egypt, and Sudan now meet had turned to desert.[118]

The cave art reminds me of when I was seven years old and my tall, dark-haired teenage cousin was looking after us, while my little cousins and I went swimming in a small stream in Carlow, Ireland, in the early 1970s. This swimming hole was formed by a small dam that slowed a stream in a field to provide drinking water for cattle. Local children would navigate cow pats for a chance to splash in the clear stream water among the shrubs and reeds. Could children of cow-herding pastoral families 6,000 years ago have done the same in the middle of the Sahara? Were their guardians on the lookout for hostile hippopotamuses that were also drawn on the cave walls? The presence of hippos suggests that the landscape at the time must have grown nutritious grass and had abundant water that was persistent enough to support populations of these giant water-loving grazers. The Sahara had been a well-watered savanna that looked like part cattle ranch and part wildlife conservation area.

---

**116** Alejandro Carrillo, in a video interview with the author on September 3, 2024, 13:21–14:18, https://youtu.be/jmrFYSHKAUE?si=imC3kOEunC46ELR6.

**117** Carrillo, 29:30–30:10.

**118** Heiko Riemer et al., "Climate, Styles and Archaeology: An Integral Approach Towards an Absolute Chronology of the Rock Art in the Libyan Desert (Eastern Sahara)," *Antiquity* 91, no. 355 (2017): 20–22, https://doi.org/10.15184/aqy.2016.252.

For a long time, scientists assumed that the Sahara was historically a dry region punctuated by wetter times, but a 2025 research study of hydrogen and carbon isotopes from leaf waxes in ocean sediments off the coast of West Africa has revealed the opposite. Saharan summer monsoons were strong enough to support savanna much farther north than today, going back 10 million years—even during the cooling and drying shift that occurred with the onset of ice ages 2.6 million years ago, as the world entered the Pleistocene.[119]

Astonishingly, between 11,000 and 5,000 years ago, there were megalakes, green plants, and thriving herds of many species of grazing animals on the land that is now the Sahara Desert. This wet period, caused by a strengthening of the West African monsoon, was one of a series of many humid periods in the past.

Although the Saharan humid period was initiated by the Milankovitch cycles that sent stronger sunlight to the region, orbital effects alone could not have produced such a wet landscape. University of Toronto physicists applied computer models to look at climate feedback effects from the simultaneous presence of green vegetation, the higher soil organic matter that would have existed under more vegetation, and the presence of megalakes that extended over 6 percent of North Africa. They found that the presence of all three was required to combine with the Milankovitch cycles to maintain strong enough monsoon rains to keep the entire North African landscape green.[120] Their models found that landscape feedback effects from a green ecosystem produced 800 mm more rainfall in the Sahel and generated enough rain in the rest of the Sahara to water a steppe grassland. No one species of animal or plant could have supercharged the productivity of the waving seas of grass dotted with abundant game and trees. Only the synergy of millions of symbiotic relationships between plentiful life forms could total a biosphere that induced climate-cooling rain.

## Yesterday's and Tomorrow's Water

Did humans speed up the desertification of the Sahara over the last 8,000 years? Overgrazing by livestock and farming between 8,000 and 4,500 years ago in the Sahara produced bare soil and caused erosion that damaged water cycles. This may have pushed the climate past tipping point faster. According to David K. Wright, who studied desertification in Northern Africa, "Evidence from the Sahara,

**119** Bryce A. Mitsunaga et al., "Fundamentally Unchanged Northwestern African Rainfall Regimes Across the Plio-Pleistocene Transition," *Science Advances* 11, no. 25 (2025): 1, https://doi.org/10.1126/sciadv.ads3149.

**120** Deepak Chandan and W. Richard Peltier, "African Humid Period Precipitation Sustained by Robust Vegetation, Soil, and Lake Feedbacks," *Geophysical Research Letters* 47, no. 21 (2020): sect. 4, para. 1, https://doi.org/10.1029/2020GL088728.

in particular, finds correlative patterns between the arrival of domesticated stock and the changeover of the landscape from high NPP (net primary productivity) to desert." [121]

Fully functioning ancient ecosystems once filled vast aquifers with the fresh water we use to grow food for humanity today. In 2007, the ancient shores of a megalake that existed 10,000 years ago in Darfur were discovered by Eman Ghoneim of the Boston University Center for Remote Sensing using radar data from space. The Northern Darfur Megalake once covered over 30,000 $km^2$ and was larger than Lake Erie on the US–Canadian border. Its well-defined shorelines reveal that lake levels were stable for long periods of time. This discovery is indisputable proof that the Sahara was once a wet, green region.[122]

The Sahara experienced 230 wet periods over the past eight million years. Pollen records indicate that during the Eemian wet period, 128,000 to 122,000 years ago, uniquely diverse savanna and open woodland covered the entire Sahara. The strength and duration of the wet period and the diversity of life that self-organized at the time may have been representative of all the previous wet periods as far back as six to eight million years ago.[123]

These greener lands of the Eemian period were kept lush and vibrant by Africa's mighty megafauna. The thriving, fully functioning ecosystems they cultivated built carbon-rich soils deep enough to store water and sustain the wetter climate. Did humans create the Sahara Desert by removing many of the animals that cycled and spread nutrients? If future humans are to have access to the vast water reserves required for agriculture, we need to restore the missing green, hydrating ecosystems that have been erased by our ancestors around the world. Can mimicking the natural behavior of megafauna by how we manage the most abundant modern megafauna, people and livestock, restore rain-inducing green ecosystems to regions where aquifers of fossil water are running lower by the minute?

---

**121** David K. Wright, "Humans as Agents in the Termination of the African Humid Period," *Frontiers in Earth Science* 5 (2017): conclusion, para 1, https://doi.org/10.3389/feart.2017.00004.

**122** E. Ghoneim and F. El-Baz, "DEM-Optical-Radar Data Integration for Palaeohydrological Mapping in the Northern Darfur, Sudan: Implication for Groundwater Exploration," *International Journal of Remote Sensing* 28, no. 22 (2007): 5009–13, https:// doi.org/10.1080/01431160701266818.

**123** Juan C. Larrasoaña et al., "Dynamics of Green Sahara Periods and Their Role in Hominin Evolution," *PLOS ONE* 8, no. 10, e76514 (2013): 9, https://doi.org/10.1371/journal.pone.0076514.

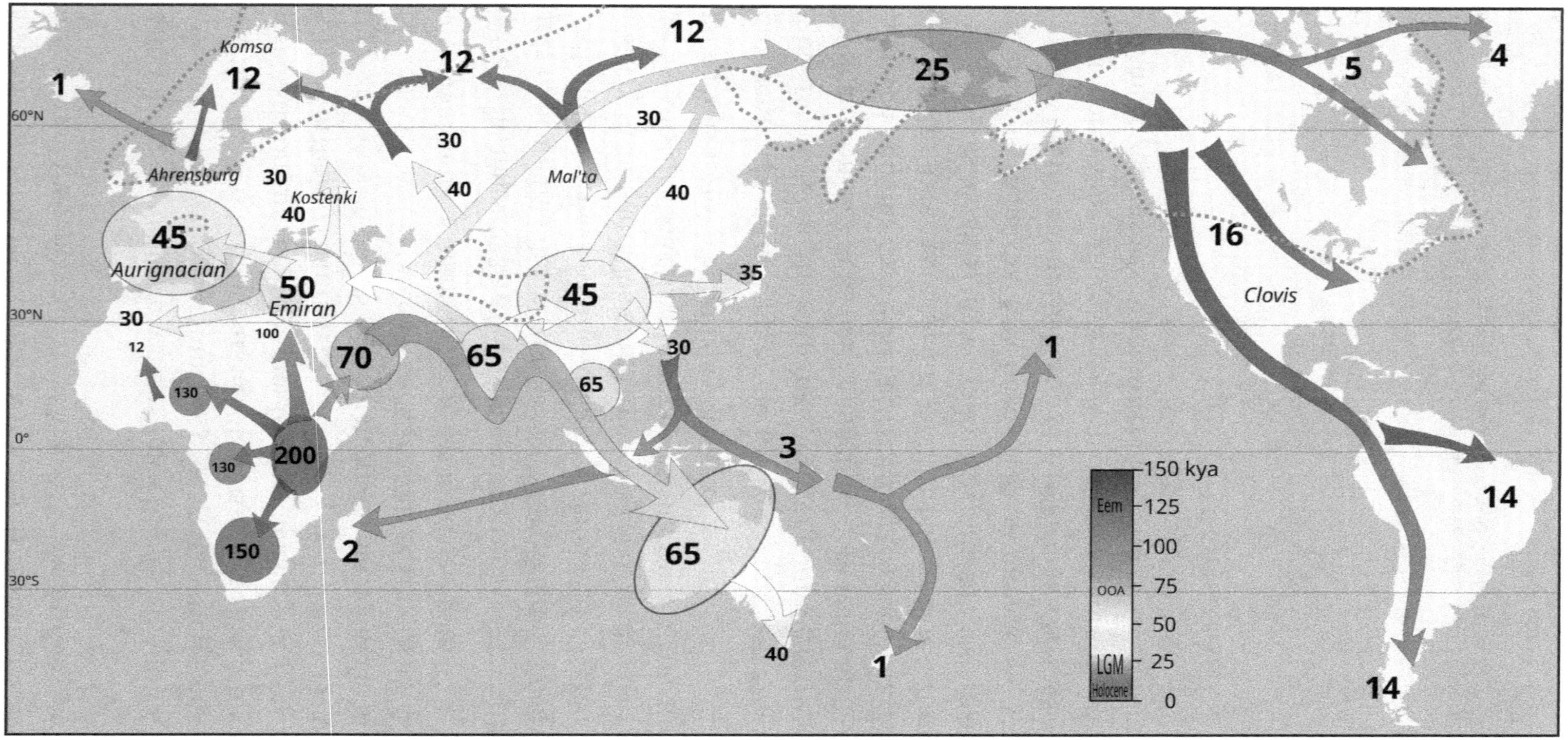


**Figure 6.2 World map of the peopling of the world. Figures are in thousands of years ago (kya).** Dotted lines are meant to indicate approximate glaciation during the Last Glacial Maximum (LGM).

Source: Wikipedia. Created by User:Dbachmann - Own work; CC BY-SA 4.0.

## Biology as a Climate Driver

Some scientists think human hunting of megafauna impacted the climate in the past by reducing landscape function past tipping points.[124] If achieved simultaneously over many continents, could deteriorating ecosystems have caused a reduction in carbon cycling, water cycling, and total photosynthesis powerful enough to warm the global climate?

A decline in megafaunal abundance accelerated between 76,000 years ago and 32,000 years ago as humans migrated out of Africa (**Figure 6.2**). Today, the biomass of all wildlife, big and small, is only 10 percent of the biomass of megafauna who once roamed the Earth.[125] No place on Earth retained all the teeming giants our ancestors used as subjects for their rock art.

An alternative theory for the extinction of so many megafauna is that climate change caused it. But a study that compared the probability of human population density and megafaunal extinction rate change found a 96 percent correlation. There was only a 63.6 percent correlation between temperature changes and megafaunal extinction.[126] This makes it hard for many scientists to imagine something other than human hunting pressure pushing megafaunal numbers past sustainable tipping points.[127, 128, 129, 130, 131] Accordingly, humans played an essential role in the extinction of the woolly mammoth.[132] Fossils have even been discovered with spear points and butchery cuts in the bones of these slaughtered giants.[133, 134]

**124** Christopher E. Doughty et al., "Biophysical Feedbacks Between the Pleistocene Megafauna Extinction and Climate: The First Human-induced Global Warming?," *Geophysical Research Letters* 37, no. 15 (2010): para. 20, https://doi. org/10.1029/2010GL043985.

**125** Bergman et al., "Worldwide Late Pleistocene and Early Holocene Population Declines in Extant Megafauna Are Associated with *Homo sapiens* Expansion Rather than Climate Change," *Nature Communications* 14:7679, no. 1 (2023): Disc. Para 4, https://doi.org/10.1038/s41467-023-43426-5.

**126** Andermann et al., "The Past and Future Human Impact on Mammalian Diversity," Conclusion, para 1.

**127** Yadvinder Malhi et al., "Megafauna and Ecosystem Function from the Pleistocene to the Anthropocene," *Proceedings of the National Academy of Sciences of the United States of America* 113, no. 4 (2016): 839–840, https://doi.org/10.1073/ pnas.1502540113.

**128** Luciano Prates and S. Ivan Perez, "Late Pleistocene South American Megafaunal Extinctions Associated with Rise of Fishtail Points and Human Population," *Nature Communications* 12, no. 1, 2175 (2021): 9, https://doi.org/10.1038/s41467-021-22506-4.

**129** Christopher Sandom et al., "Global Late Quaternary Megafauna Extinctions Linked to Humans, not Climate Change," *Proceedings of the Royal Society B: Biological Sciences* 281, no. 1787 (2014): sect. 4, para. 1, 20133254. https://doi. org/10.1098/rspb.2013.3254.

**130** Felisa A. Smith et al. "Methane Emissions from Extinct Megafauna." *Nature Geoscience* 3, no. 6 (2010): 375, https://doi. org/10.1038/ngeo877.

**131** Van der Kaars et al., "Humans Rather than Climate the Primary Cause of Pleistocene Megafaunal Extinction in Australia," Disc., para. 4.

**132** Damien A. Fordham et al., "Process-Explicit Models Reveal Pathway to Extinction for Woolly Mammoth Using Pattern-Oriented Validation," *Ecology Letters* 25, no. 1 (2022): 134, https://doi.org/10.1111/ele.13911.

**133** Eileen Johnson and Leland C. Bement, "Bison Butchery at Cooper, a Folsom Site on the Southern Plains," *Journal of Archaeological Science* 36, no. 7 (2009): figs. 4 and 5, https://doi.org/10.1016/j.jas.2009.02.007.

**134** Michael R. Waters et al., "Late Pleistocene Horse and Camel Hunting at the Southern Margin of the Ice-Free Corridor: Reassessing the Age of Wally's Beach, Canada," *Proceedings of the National Academy of Sciences of the United States of America* 112, no. 14 (2015): 352, https://doi.org/10.1073/pnas.1420650112.

## The Great Floods

During the last glaciation, parts of the far north were not covered in ice sheets. A massive region encompassing the mammoth steppe stretched across Northern Europe, large swathes of Russia, across the exposed land bridge into Alaska, and over to the Yukon region of Canada. When my European ancestors first reached western Europe by 40,000 years ago, they encountered landscapes that were frigid in winter. But, in summer the sunny steppe grassland was awash in color from flowering plants and the bright green of grasses kept vegetative and fertilized by herds of giants traversing the landscape.

It was colder but drier than today, so less snow fell on the steppe grasslands in winter. Huge grazing animals could easily dig down through snow to access frozen grass throughout the long winters, just like I witnessed our cows and wild deer do at Raindance Farm. The year-round presence of large game animals like mammoth, reindeer, musk ox, woolly rhinoceroses, and aurochs enticed my ancestors to innovate so they could exploit this abundant source of energy year-round. People had learned to sew leather garments that formed to their bodies to keep cold air out and they could use mammoth bones and skins to build shelter when caves were unavailable. Fire to gather around on cold winter nights made life possible for growing populations of people.

Over time, effective hunting technology and social cooperation came at the expense of the largest grazing animals that were hunted until they only existed in fireside monster stories. As large grazing animals became harder to find over the vastness of space and time, the grasslands they once kept open became choked with brush that made it harder to see game and predators like wolves and bears. Hunting strategies had to change or people had to move. The vast carpets of poppies and other flowering plants that children once ran amongst became shaded out over time by willow shrubs and small birch trees.

Ice core samples reveal several sharp rises in sea level in the distant past that occurred soon after rises in $CO_2$ levels in the atmosphere. The timing of these spikes in global temperature and sea level is about one or two thousand years after consecutive cohorts of megafauna species were eliminated on the largest biome on the planet at the time, the mammoth steppe, and in North and South America. In the Yukon, mammoth, mastodon, steppe bison (*Bison priscus*), and two species of horses (*Equus scotti* and *Haringtonhippus francisci*) were eradicated starting around 22,000 years ago. This coincides with the oldest human footprints we know of that were made in North America at White Sands, New Mexico, between 23,00 and 21,000 years ago. Mammoth and mastodon species were the first to suffer a steep decline, followed by the steppe bison, and then horses.

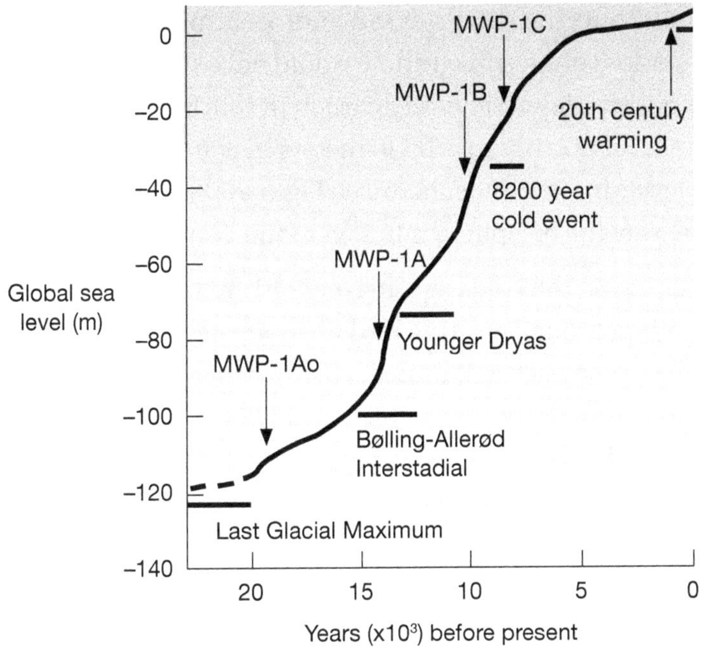


**Figure 6.3 Generalized sea level rise since the last ice age showing several meltwater pulses (MWP).** MWP-1A0, c. 19,600 to 18,800 years ago; MWP-1A, 14,600 to 13,800 years ago; MWP-1B, 11,000 to 8,800 years ago; and MWP-1C, c. 8,200 to 7,600 years ago.

Adapted from Vivien Gornitz, NASA (2012).

Could the removal of keystone species have initiated a cascading destructive effect on ecosystem productivity that released $CO_2$ into the atmosphere and warmed the Earth?

A great flood, referred to by paleoclimatologists as meltwater pulse 1Ao, occurred between 19,600 and 18,800 years ago (**Figure 6.3**). A sedimentary ancient DNA (sedaDNA) sample from the Yukon dated between 20,300 years ago and 19,000 years ago shows a sharp decline in the number of woolly mammoths, American horses, and bison occurred just before this spike in temperatures and sea levels.[135] Hunting pressure by humans on mammoths and bison would have reduced the animals' effectiveness as ecosystem engineers to maintain the once highly productive mosaic ecosystem.[136]

Prior to the sharp decline in mammoths around 20,000 years ago, analysis of seda-DNA has revealed, the Yukon had much richer diversity and abundance of flowering plants and grasses. If seda-DNA reflects absolute abundance, then the sheer numbers of these flowering plants were greatly diminished

**135** Tyler J. Murchie et al., "Collapse of the Mammoth-Steppe in Central Yukon as Revealed by Ancient Environmental DNA," *Nature Communications* 12, no. 1 (2021): 5, fig. 2, https://doi.org/10.1038/s41467-021-27439-6.

**136** S.A. Zimov et al., "Mammoth Steppe: A High-Productivity Phenomenon," *Quaternary Science Reviews* 57 (2012): 27, https:// doi.org/10.1016/j.quascirev.2012.10.005.

and never recovered upon the disappearance of vast numbers of the largest grazers.[137]Biodiversity loss of such magnitude would have slowed both ecosystem productivity and opportunities to sequester carbon in soil, leaving more carbon in the atmosphere. Sea levels swiftly rose by 10 meters in 800 years in lockstep with slightly higher $CO_2$ levels in the atmosphere (see **Figure 6.3**). [138] This is a profound manifestation of the way the biosphere is linked to the atmosphere.

What was happening on the landscape in New York State all those years ago as glaciers retreated? When ground-nesting birds and Karner blue butterflies, who love grasses and forbs, came back to Raindance Farm, what were they looking for? The first human hunters to arrive in the Northeast US would have been thrilled with the game opportunities that abounded. All the characters of the mammoth steppe and more, including giant sloths, had been thriving south of the reach of glaciers for millions of years.

As land was exposed with the retreat of ice age glaciers, animals migrated north to co-create thriving tundra and then steppe grassland ecosystems in New York State. After humans depleted the abundant game on the mammoth steppe, they had their sights set southward on the highly productive ecosystems that had evolved in the US, Mexico, and Central and South America.

Examination of the abundance of ancient megafaunal dung spores from soil core samples shows that the median timing of megafaunal decline at several sites in the Eastern U.S. was between 16,000 and 13,800 years ago, and at most sites it was immediately followed by a shift from the park-like savanna that was the former ecosystem at Raindance Farm, to more closed-canopy deciduous forest.

Megafaunal populations south of the glaciers in the Northeast and Central US crashed before the flowers on the savannas were replaced by drab shrubbery. This supports the megafaunal release hypothesis that megafaunal extinction removed natural grazing pressure on trees so more of them grew, and this changed the shape of their ecosystems—not climate change. But there is also some truth to the climate change hypothesis because as grassland and steppe ecosystems lost their megafaunal gardeners and collapsed into woody, less productive landscapes, lower albedo, the influence of soil carbon losses and slowed soil carbon sequestration would indeed lift atmospheric $CO_2$ levels and warm climate. This made it harder for the remaining megafauna to persist.

**137** Murchie et al., "Collapse of the Mammoth-Steppe," 7, fig 4.

**138** Vivien Gornitz, "The Great Ice Meltdown and Rising Seas: Lessons for Tomorrow," NASA Goddard Institute of Space (2012), accessed July 26, 2025, http://350.me.uk/TR/Hansen/NASA%20GISS-%20The%20Great%20Ice%20Meltdown%20and%20Rising%20Seas-%20Lessons%20for%20Tomorrow.pdf

## The Collapse of the Mammoth Steppe

After mammoths became functionally extinct in the Yukon 19,000 years ago, numbers of faster-reproducing species like horses and bison rebounded and helped keep trees and shrubs from invading some of the mammoth steppe for several thousand years. But they too eventually succumbed to human overhunting. Over the period between 22,400 years ago and 12,700 years ago, seda-DNA evidence reveals the lowest abundance of horse, mammoth, and bison in the Yukon occurred between 16,800 years ago and 14,700 years ago. This coincided with a rise in atmospheric $CO_2$, likely a result of the declining ecosystem productivity created by the removal of these fertilizer makers and spreaders. The smoking-gun signal of this ecosystem restructuring is seen in the sharp drop in the ratio of carbon-13 isotopes ($^{13}C$) to carbon-12 isotopes ($^{12}C$) measured in $CO_2$ from bubbles in ice core samples from Antarctica between 17,600 and 15,500 years ago (**Figure 6.4**).

Plants prefer to absorb $^{12}C$ during photosynthesis more than the heavier $^{13}C$, so the decay of plant-derived products like soil carbon or the burning of trees (and even fossil fuels that once were plants) would release $CO_2$ into the air that is richer in $^{12}C$ compared to $^{13}C$, lowering the $^{13}C:CO_2$ ratio. The large anomalies in the makeup of atmospheric carbon centered at 16,300 years ago, and again at 12,300 years ago, reflect rapid releases of terrestrial carbon to the atmosphere in just a few hundred years.[139] These data support the idea that the killing of so many megafauna dramatically lowered land productivity on the North American continent and later in South America. Meanwhile, total atmospheric $CO_2$ rose as the transition from grassland to forest released soil carbon and exerted warming forcings on global climate.

It is no surprise that Vostok ice core evidence reveals the most rapid global sea level rise event since the coldest part of the glaciation occurred within 1,000 years of the initiation of rapid dung spore decline (indicating megafaunal decline) in North America and the abrupt $\delta^{13}C:^{12}C$ anomaly between 17,600 years ago and 15,500 years ago. Meltwater pulse 1A occurred between 14,700 and 13,500 years ago as $CO_2$ jumped 10 parts per million (ppm), causing a global mean sea level rise of 17.9 m in 500 years.[140] If this occurred today, Hell's Kitchen in Manhattan, my birthplace, would be under 5 m of water.

**139** Thomas K. Bauska et al., "Carbon Isotopes Characterize Rapid Changes in Atmospheric Carbon Dioxide During the Last Deglaciation," *Proceedings of the National Academy of Sciences of the United States of America* 113, no. 13 (2016): 3469, https://doi.org/10.1073/pnas.1513868113.

**140** Yucheng Lin et al., "A Reconciled Solution of Meltwater Pulse 1A Sources Using Sea-Level Fingerprinting," *Nature Communications* 12, no. 1 (2021): 2015, abstract, , https://doi.org/10.1038/s41467-021-21990-y.

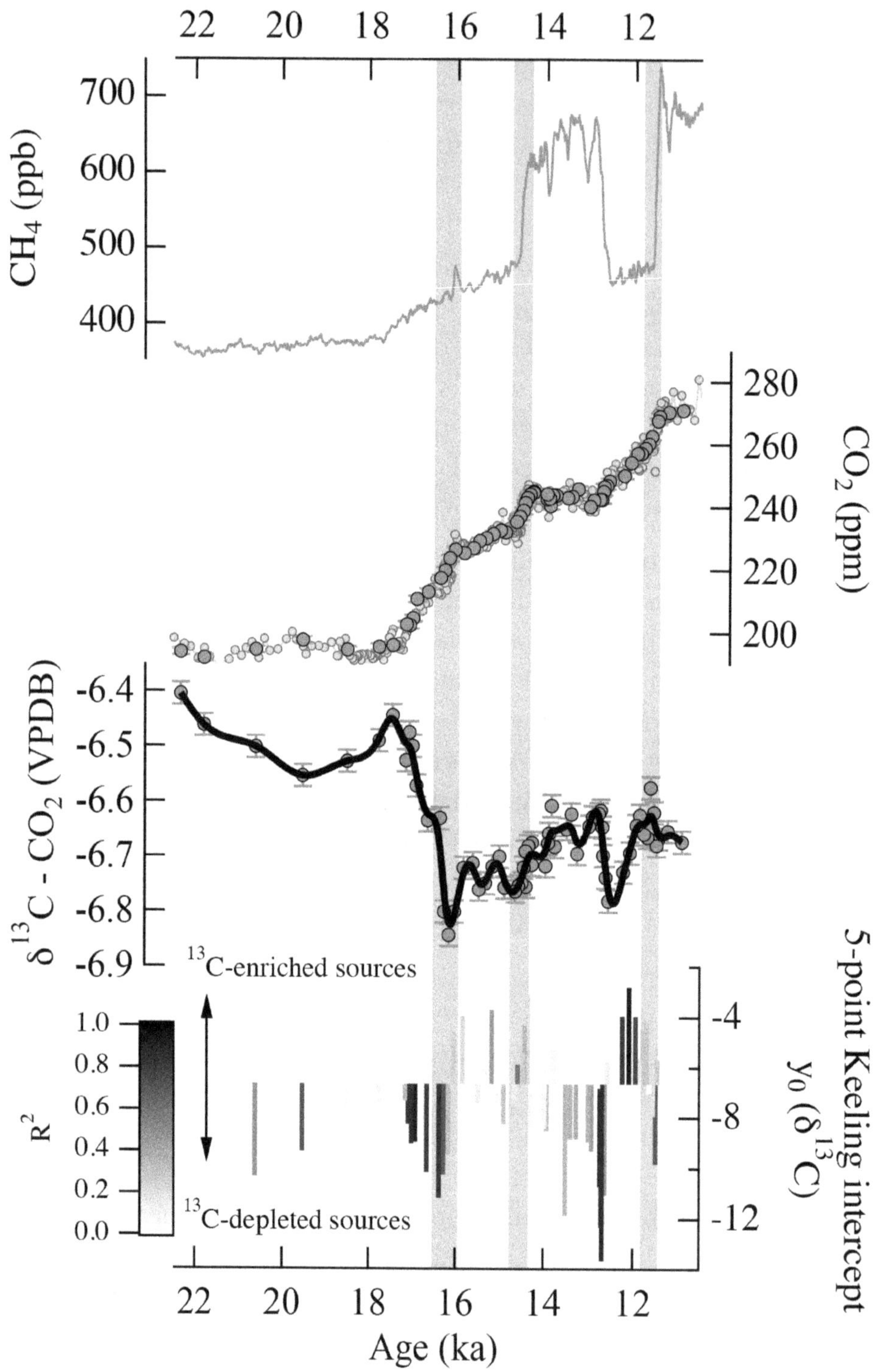


**Figure 6.4 Carbon cycle changes of the last deglaciation.** Vertical bars indicate intervals of rapid $CO_2$ rise identified in ice cores from the West Antarctic Ice Sheet.

Source: Bauska et al., 2017.

During and after meltwater pulse 1A, big changes were occurring to the landscape in North America. At Appleman Lake, Indiana, a core sample 11.3 meters deep revealed that the abundant spores associated with megafauna dung fell below 2 percent of the total pollen and spore count 13,000 years ago and never rebounded. There, and in New York State, scientists recorded more pollen of palatable shade-tolerant trees like hop-hornbeam, elm, spruce, and larch that preferentially drew down $\delta^{12}C$ from the atmosphere for a short time, slowing the $CO_2$ rise but raising the Carbon 13 isotope ratio briefly as they established. But these forests lacked the biodiversity and soil building ability of former open savanna and are less productive in the long run. The landscapes where humans once could gaze far into the distance as they searched for game among ground-nesting birds and butterflies, became blocked with dark, closed-canopy forest. These were more likely to burn hot if a wildfire started because biomass uneaten by megafauna built up in the under story and tree canopies were no longer spaced apart.

The sudden jumps in atmospheric $CO_2$ that occurred in just a few hundred years during meltwater pulse 1A, and again at 12,800 years ago, cannot be explained by ocean current or sea temperature changes alone. There likely was a land-based source of the enrichment of photosynthetically derived carbon at the time.[141] Since biology was so much more abundant in the Pleistocene, it was a stronger driver of climate than today, with the carbon and water cycles as the levers.

## The Mammoth Steppe Rallies

When the grassy mammoth steppe ecosystem finally collapsed between 13,500 and 10,000 years ago, it was replaced by early succession trees, primarily willow (*Salix* spp.) and dwarf birch (*Betula nana*).[142] At first, the encroachment of woody shrubs and small trees rapidly drew down immense quantities of $CO_2$ which had a cooling effect on climate. But later, the less productive forests would leak soil carbon that had been sequestered by the former grasslands into the air.

The net effect of wild, giant, grazing and browsing animals on landscapes is a build-up of carbon-rich soil and peat and net removal of $CO_2$ from the atmosphere. Interestingly, the temporary cooling of the Younger Dryas in the Northern Hemisphere, roughly between 12,900 and 11,700 years ago, was immediately

**141** Bauska et al., *Carbon Isotopes Characterize Rapid Changes in Atmospheric Carbon Dioxide During the Last Deglaciation* 3468.

**142** Murchie et al., "Collapse of the Mammoth-Steppe," 9.

preceded by an uptick in South American megafauna numbers between 15,500 and 13,500 thousand years ago (**Figure 6.7**) and bison, horse, forb, and grass seda-DNA on the mammoth steppe.[143]

As surging numbers of South American megafauna like giant ground sloths devoured more trees and shrubs to let light down to ground level, the area of grass expanded within landscape mosaics. The drawdown of $CO_2$ from the soil building effect of South American megafauna and expansion of woody shrubs in North America would explain why the carbon 13 ratio increased between 15,500 and 13,500 years ago (**Figure 6.4** and **6.5**). The net result was the subsequent cooling period of the Younger Dryas.

This period may have been the last days of the productive ecosystem that had helped keep the Earth cooler during the ice age. But it didn't last long. As much more human seda-DNA was observed in the fossil record 13,000 years ago, the numbers of horses and bison crashed. Terrestrial carbon flooded the atmosphere from both the collapse of the mammoth steppe and the abrupt megafaunal extinctions occurring in South America as evidenced by the sharp drop in the Carbon 13 ratio. (**Figure 6.4**). But this time, the rapid widespread flush of woody growth as forest was released from grazing pressure was powerful enough to draw down enough $CO_2$ to keep temperature lower during the Younger Dryas.

While meltwater pulse 1A occurred at a time when the warming effect of solar insolation was increasing, its severity was amplified by the relentless human slaughter of megafauna in the Americas, Europe, and Asia on continental scales (**Figure 6.5**). The demise of megafauna was a planet-wide phenomenon that would have changed the face of the planet as ecosystems shifted to new regimes and species make-up.[144, 145] I can't think of anything more powerful than an 18-m flood to demonstrate the value in restoring fully functioning ecosystems to prevent this from happening again.

**143** Murchie et al., "Collapse of the Mammoth-Steppe," 5, fig. 2.

**144** Doughty et al., "Biophysical Feedbacks," sect. 3, para. 6.

**145** Pearce et al., "Higher Abundance," sect. 4, para. 6.

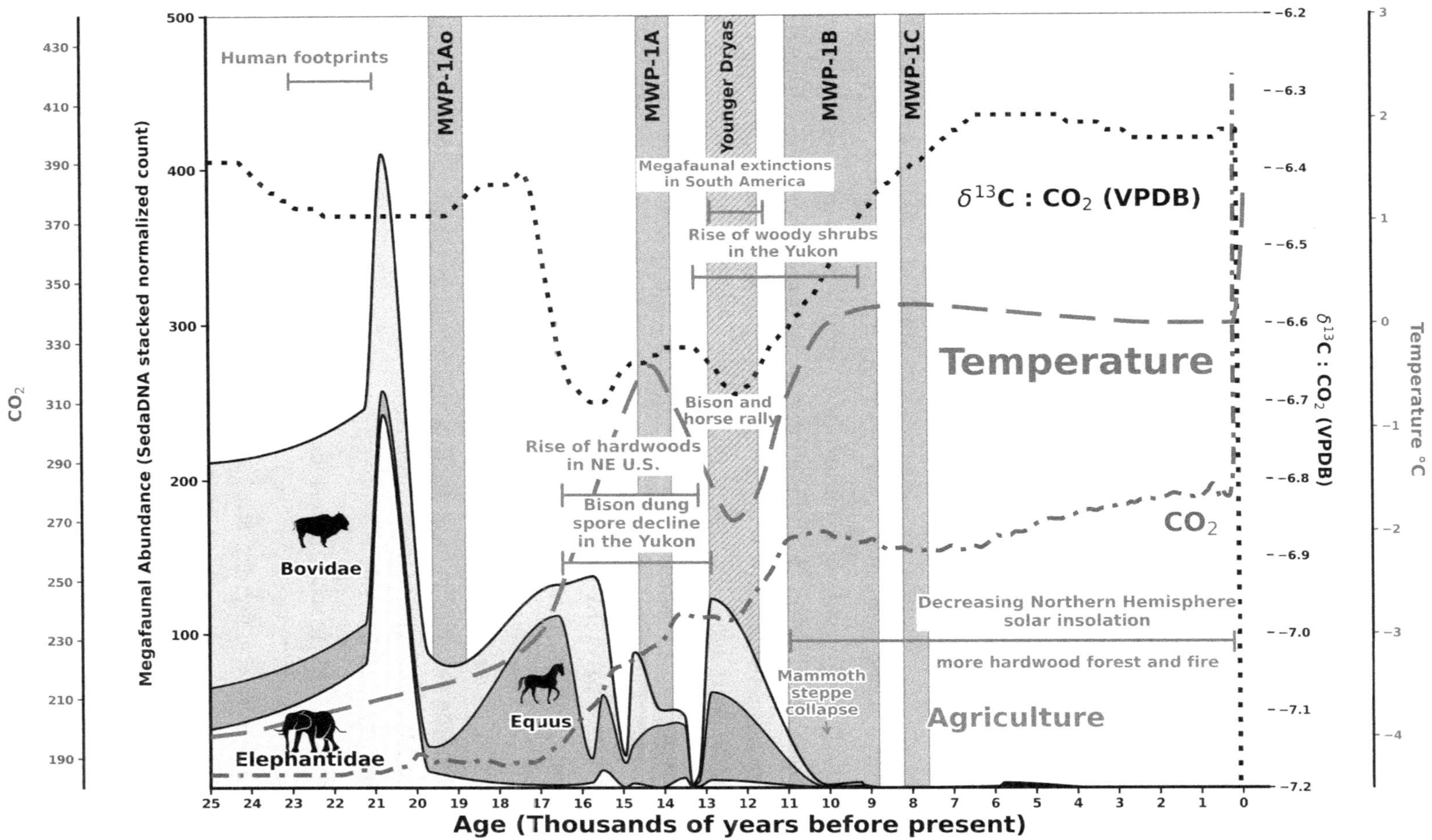


**Figure 6.5. The biosphere is linked to the atmosphere via the carbon cycle.** Concept by Siobhan Griffin. Illustrated by Mohamed Ahmed Abdelfattah

## Horses on the Menu and More Great Floods

The ramifications of human hunting continued to impact the global climate with the collapse of the final remnants of the mammoth steppe ecosystem. Meltwater pulse 1B produced a 7.5-m sea level rise over 160 years, centered around 11,000 years ago. Horses in the far north of the Americas had survived at populations high enough to maintain the last vestiges of climate-cooling steppe grassland until about 10,000 years ago.[146] Then the horses and grasses disappeared together, allowing a less diverse and less productive ecosystem of mosses, willows, and pines to fully establish. Many of the colorful flowers were gone, replaced by a dark green monotony of dwarf trees on the landscape that absorbed more of the sun's heat and warmed the vast northern reaches of the planet.

Other impacts arising from the final collapse of the mammoth steppe would have contributed to global warming. Charcoal from forest fires increased in the fossil record after megafaunal grazing pressure stopped resisting tree growth, especially after 11,000 years ago. Remember forests only sequester significant carbon in the first 100 years of establishment. In the long run, more frequent fire would have added both $CO_2$ and methane to the atmosphere, preventing the global cooling that normally would have occurred as solar insolation began to decrease 11,000 years ago.

Additionally, with fewer hooved grazing animals to trample snow, large pulses of the planet-warming greenhouse gas methane was released from melting permafrost during MWP 1A and just prior to meltwater pulse 1B. Herds of larger grazing animals on the grassy steppe, such as mammoth, horse, musk ox, saiga antelope, and caribou, once trampled insulating snow and allowed cold winter air to meet the soil. As a result, soil froze deeper underground. This animal impact helped prevent the permafrost from melting and releasing the potent greenhouse gas methane. Methane release from permafrost soils in the distant past have been linked to abrupt spikes in climate temperature of up to 6 degrees Celsius.[147]

We witnessed the insulating effect of snow at Raindance Farm, where sometimes our annual rye cover crop would grow under the snow in below-freezing conditions with an insulating layer of snow above. The soil would thaw from the bottom up under the blanket of snow. Meanwhile, on our cow lanes where the cows trampled down the snow, the frost would penetrate 1-m-deep during our cold snaps, which sometimes were as cold as –35° Celsius (–31° F).

**146** William T. Taylor, *Hoof Beats: How Horses Shaped Human History* (University of California Press, 2024), 26.

**147** Robert M. DeConto et al., "Past Extreme Warming Events Linked to Massive Carbon Release from Thawing Permafrost," Nature 484, no. 7392 (2012): fig. 2. https://doi.org/10.1038/nature10929.

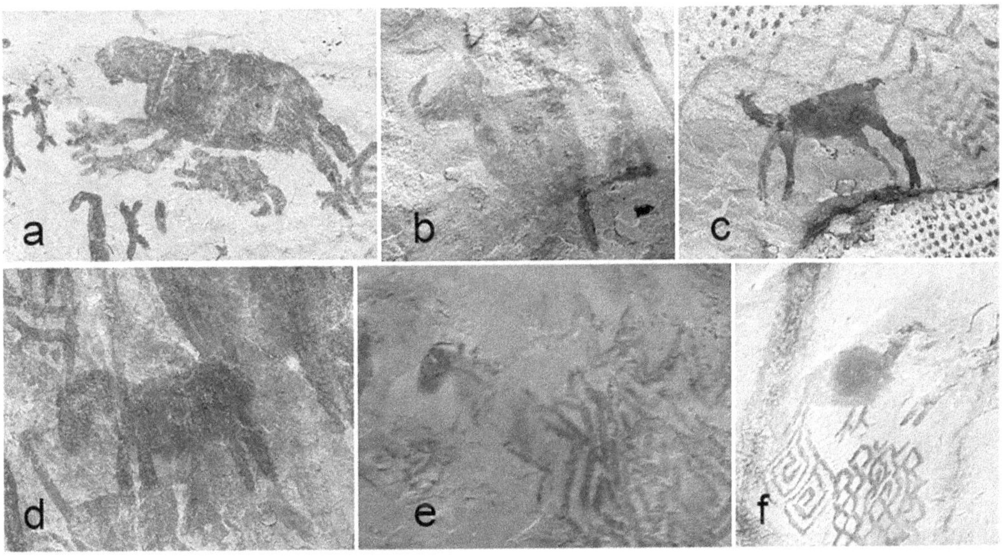


**Figure 6.6 Potential ice age megafauna displayed in the pictographs of La Serranía La Lindosa, Colombia, including: a. giant sloth, b. mastodon, c. camelid (Paleolama?); d and e. horse; f. long-neck, three-toed ungulate with trunk (Xenorhinotherium, Macrauchenia?)**

Adapted from Gaspar Morcote-Ríos et al., 2021.

The same effect happens under driveways in cold climates when people remove the snow. The frost will go much deeper than under the adjacent snow-covered lawn.

## New Meat-eaters Arrive in South America

Evidence from ice cores in Antarctica, conversely, show an increase in levels of atmospheric $CO_2$ and southern hemisphere warming during the Younger Dryas, when climate got colder in the northern hemisphere.[148] This could be expected, as human populations were making their mark. Homo sapiens had crossed into South America by 15,500 years ago and spread quickly throughout the continent.[149] They encountered abundant game animals, like those depicted in artwork on the cliffs at La Serranía de la Lindosa in northwest Colombia (**Figure 6.6**).[150] South America had the most megafauna species of any continent. Gone forever now are towering mastodon, giant sloths, and large camelids, whose fertilizing effect may have spread limiting nutrients like phosphorous sufficiently to keep the pampas and savannas vigorously growing.[151]

---

**148** Vera Markgraf, "Younger Dryas in Southern South America?," Boreas 20, no. 1 (1991): 63, https://doi. org/10.1111/j.1502-3885.1991.tb00459.x.

**149** Luciano Prates et al., "Rapid Radiation of Humans in South America After the Last Glacial Maximum: A Radiocarbon-Based Study," *PLOS ONE* 15, no. 7 (2020): Disc., para. 1, https://doi.org/10.1371/journal.pone.0236023.

**150** José Iriarte et al., "Ice Age Megafauna Rock Art in the Colombian Amazon?," *Philosophical Transactions of the Royal Society B: Biological Sciences* 377, no. 1849 (2022): 4, fig. 2, https://doi.org/10.1098/rstb.2020.0496.

**151** Wolf et al., "Lateral Diffusion of Nutrients," 9.

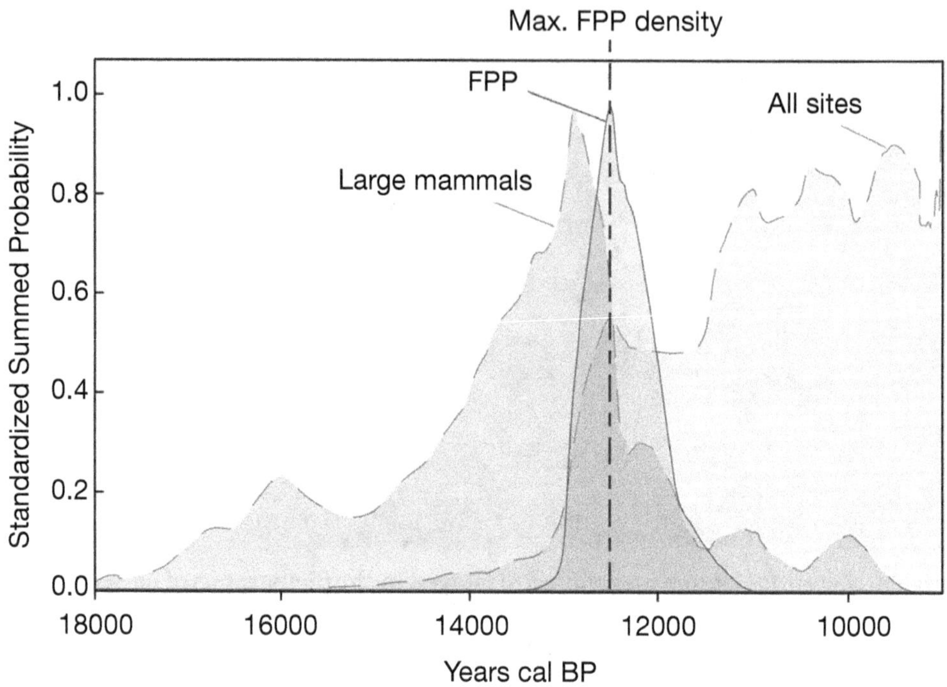


**Figure 6.7 Change in the density of evidence of megafauna and human hunting technology (fishtail projectile points or FPP).** Grey shaded area is evidence of archeological sites/human presence.

Source: Prates and Perez, 2021.

South American megafauna were hit hard by human predators, as evidenced by the strong correlation between the arrival of humans in South America and the disappearance of megafauna in fossil evidence. The sudden widespread use of fluted fishtail spear point technology 13,000 years ago gave humans access to vast amounts of energy in the form of megafaunal animal fat that fueled rapid human population growth. This was immediately followed by a steep decline in megafauna numbers throughout the continent, starting 12,900 years ago.[152] Remember, an unusual proportion of $CO_2$ rich in $^{12}C$ from land sources was flooding the atmosphere at the same time (**Figure 6.4**)—the smoking-gun signature of megafaunal extinction in South America.

Archaeologists in Argentina, Luciano Prates and Ivan Perez, documented that the close timing of widespread fishtail projectile point use, human population escalation, and a rapid crash in megafaunal numbers occurred on almost a continent-wide scale, as shown in **Figure 6.7.**[153] They reasoned that climate

**152** Prates and Perez, "Late Pleistocene South American Megafaunal Extinctions," 3.

**153** Prates and Perez, "Late Pleistocene South American Megafaunal Extinctions," 3, fig. 1.

change was not likely responsible for the South American extinctions because the megafauna populations plummeted well before pampas and Patagonian grasslands reverted to forest. They observed that "the shrinking of the grassland steppes—the most favorable environmental conditions for large mammals—from southern South America occurred between 12,400 and 11,500 years ago, 1400–500 years after the megafaunal decline (12,900 years ago) in South Patagonia; in the Pampas the change seems to have occurred even later."[154]

This time lag puts the timing of diminishing pampas squarely at the same time as meltwater pulse 1B, centered around 11,000 years ago. The area of grassland in Patagonia and on the Pampas was also shrinking due to inundation from the rising seas, and this would also have reduced the efficiency of grasslands to sequester carbon. The extinctions of megafaunal ecosystem engineers would have caused sharp reductions in pampas and forest productivity that released the additional $CO_2$ seen in Vostok ice cores. This would have had knock-on effects on the regional climate that would have amplified meltwater pulse 1B; such is the interconnectedness of the biosphere, geology, and climate.

## The Climate Impact of Agriculture

Meltwater pulse 1C occurred 8,000 years ago as people began to practice agriculture to replace their formerly abundant food sources like wild game and the diversity of plants associated with it. This time, the rise in $CO_2$ and global temperatures produced a sea level rise of 6.5 m in less than 140 years. This supports William Ruddiman's suggestion that degrading soils from farming and pastoralism released enough carbon and methane into the air to trigger a global warming event. The warming influence of agriculture continued even as solar insolation at the northern hemisphere started to weaken since meltwater pulse 1C, and the planet normally would have cooled.

If humans possess the agency to warm global climate by how we manage land and animals, why can we not have the opposite impact? This was the hope I saw as my pastures stayed green longer in dry weather at Raindance Farm. Our soil tests confirmed what our green pastures were telling us: We were pulling atmospheric carbon from the air and putting it into the soil in the form of soil carbon. We can begin to cool our planet by greening our landscapes and rebooting the water cycle, but we will need help from our fellow earthlings who possess nutrient-cycling superpowers—large grazing animals.

**154** Prates and Perez, "Late Pleistocene South American Megafaunal Extinctions," 7.

During past great floods, which are recent enough to be documented in stories and holy books, just an 80-ppm rise in atmospheric $CO_2$ resulted in a sea level rise of 120 m. Today, we have already increased atmospheric $CO_2$ by 140 ppm since the start of the Industrial Revolution, with rapid burning of fossil fuels and land use changes. With 267 million people living at less than 2 m above sea level,[155] our global climate trajectory can no longer be ignored. Climate volatility is well underway with more frequent and severe floods, droughts, and storms. According to NASA's Earthdata website, sea levels globally are rising at a rate of 4.5 mm per year, and that rate has doubled over the past 30 years. On top of that, some coastal land is subsiding by between 2 mm and 20 mm per year, so do not expect our coastal cities to resemble anything like today's by 2200 if we continue business as usual.

But sea-level rise, like most misery, is optional for humankind. Although the simultaneous degradation of ecosystems in many places warmed the Earth's climate in the past, global policies today to help people restore landscapes can restart planet-cooling water cycles. I truly believe that nature will respond by bringing back a cornucopia of life in abundance. That will be the first sign that a better, kinder future for your children and grandchildren is in store.

**155** A. Hooijer and R. Vernimmen, "Global LiDAR Land Elevation Data Reveal Greatest Sea-Level Rise Vulnerability in the Tropics," *Nature Communications* 12, no. 1 (2021): 2, https://doi.org/10.1038/s41467-021-23810-9.

## Chapter 7

# The Start of the Anthropocene

There were human-initiated megafaunal extinctions far deeper in time than those perpetrated in the Americas. These, too, impacted global climate, pushing the first geochemical evidence of the Anthropocene back to 64,000 years ago. If citizens of Earth can learn how we as a species, even at much lower numbers, had the means to warm our planet through land management, it opens the door to a commonsense climate solution that can be applied widely enough in space and time to be an effective solution to climate heating. Excess greenhouse gases will not disappear even when fossil fuels are used up. The ocean reservoir of $CO_2$ that will replenish our atmosphere for hundreds of years or more will need to be dealt with on an ongoing basis. Methane will continue to be released as long as we procrastinate in addressing climate heating and allow permafrost to melt.

The good news is that consumers everywhere can support farmers who are committed to restoring climate-stabilizing biodiversity on our foodscapes and seascapes. We citizens can urge policymakers to stop encouraging the bulldozing of nature, which serves as the foundation of all economies. Instead, citizens can insist on wise carrot-and-stick policy that will jumpstart innovation in natural economies that can sustain our progeny going forward for thousands of years. The incredible story of the powerful agency humans possess to change global climate with our management of nature starts in the arid heart of Australia.

## Australia: Megalania and Other Monsters

The remnants of Megalake Frome can be found in a desertified area of South Australia, north of Adelaide. Between 110,000 and 88,000 years ago, this megalake was full and overflowed into Kati Thanda-Lake Eyre through the Warrawoocara Channel. Forty-eight thousand years ago, the lakes were so large that water covered 25,000 $km^2$ and contained 285–400 $km^3$ of water.[156] This water was fed by

156 Tim J. Cohen et al., "Hydrological Transformation Coincided with Megafaunal Extinction in Central Australia," *Geology* 43, no. 3 (2015): 2, https://doi.org/10.1130/G36346.1.

rivers draining the Kati Thanda-Lake Eyre catchment, an area of approximately 1 million $km^2$ (an area larger than that of Italy, Poland, and Germany combined).

The landscape teemed with Australia's unique and sometimes nightmarish animals. Many kinds of reptiles plied ancient Australian waterways, including *Meiolania platyceps*, a 2.5-m-long turtle with a horned head and spiked tail. Some fish weighed as much as 65 kg. Fossils that were laid down before human arrival on the Nullarbor Plain in the southwest corner of South Australia include large tree trunks, many species of tree kangaroo, and crocodiles, which suggests a previously hydrated ancient landscape that was probably a mosaic of savanna, forest, swamps, rivers, and lakes.

The productivity of the extensive Kati Thanda-Lake Eyre catchment rivaled that of tropical wetlands that are known for their flourishing wildlife. Paleontologist Steve Webb, who collated much of the fossil evidence around the now usually dry salt flats at Kati Thanda-Lake Eyre, found the fossil evidence of abundant life he uncovered "almost too incredulous to contemplate."[157] This highly evolved diversity of life is evidence that the ecosystem had sustained itself over millions of years.

There is both fossil and cultural evidence of the extreme abundance of life on ancient Australian plains. The First Nations Barngarla people's use of the plural to describe groups of emus is intriguing. They use *warraidyarri* for a group of emus (up to 20 individuals) and *warraidyailyarranha* for a massive gathering of emus (100+ birds).[158] Could there have been a time in our human past when simply using a plural noun was inadequate to describe the magnitude of the abundance of other species that we shared our landscapes with?

Isotopic analysis of the eggshells of fossilized giant birds *Genyornis newtoni and Dromaius novaehollandiae*, as well as wombat fossils, reveal diets richer in nutritious grasses prior to a large extinction event 45,000 years ago. This has led some scientists to believe that a far more diverse landscape supported the abundance of animal and reptilian species represented in fossils from Australia's now-arid interior.[159] What happened to this lost world and the animals that had survived for millions of years prior to the arrival of the first humans?

**157** Steve Webb, "Palaeotrophic Reconstruction and Climatic Forcing of Mega-Lake Eyre in Late Quaternary Central Australia: A Review," *Boreas* 39, no. 2 (2010): 323, https://doi.org/10.1111/j.1502-3885.2009.00120.x.

**158** Emmalene Richards and Ghil'ad Zuckermann, *Mangiri Yarda* (*Healthy Country: Barngarla Wellbeing and Nature*), Part 1 (Revivalistics Press, 2021), 9.

**159** Gifford H. Miller et al., "Ecosystem Collapse in Pleistocene Australia and a Human Role in Megafaunal Extinction," Science 309, no. 5732 (2005): 290, https://doi.org/10.1126/science.1111288.

## Animals Knit Australia's Green Cloak

At Raindance Farm, our landscape had the inherent capacity to be more productive if we added fertilizer because we were missing the faunal diversity that once would have turbocharged nutrient cycling. We purchased organic chicken litter as a source of nitrogen fertilizer to mimic the naturally fertilized grasslands of the past. I reasoned that agroecosystems with only a few species of animals like ours would benefit from the fertility of other species. Our small flock of 80 chickens was not big enough to mimic the nitrogen cycling ability of the teeming flocks of birds that were present in the past. We also purchased rock phosphate as a phosphorus source and lime for calcium to replace the nutrients that were exported from the farm when we sold nutrient-dense milk and meat.

Australia's ecosystems were uniquely fragile even before humans decreased ecosystem function because Australian soils are ancient and have not benefited from the deposition of minerals released by glaciers, as occurred in the northern hemisphere and South America during recent ice ages. This makes most Australian soils deficient in nutrients like carbon and phosphorus.[160]

Nature had an elegant workaround for this problem. As animals ate plants and were subsequently eaten themselves by predators, nutrients passed up the food chain. The interconnected dynamics between many more species than we have today, each existing in unimaginable abundance, acted to spread plant growth-limiting nutrients like nitrogen and phosphorus laterally across the landscape from nutrient hot spots near the coast. This improved Australia's ability to harvest the strong sun at her latitude, making the continent far more productive than we can ever imagine, as was the case before humans impacted other landscapes across the planet.

*Diprotodon optatum* were herbivorous marsupials the size of a rhino that would have wandered widely, or possibly migrated, to find the large quantities of food they required before their extinction 45,000 years ago. *D. optatum* fossils as young as 47,000 years old discovered near Kati Thanda-Lake Eyre and elsewhere indicate that herds were dispersed widely around the now-arid Australian interior. Trains of moving herds transported critical nutrients in the form of dung and carcasses from fertile coastal areas to less fertile interior lands as they traversed the continent looking for food.

160 David J. Eldridge et al., "Australian Dryland Soils Are Acidic and Nutrient-Depleted, and Have Unique Microbial Communities Compared with Other Drylands," *Journal of Biogeography* 45, no. 12 (2018): 2807, https://doi.org/10.1111/jbi.13456.

Megafauna are far more effective than smaller animals at spreading nutrients across landscapes because they travel relatively long distances, their food takes longer to digest, and they make more dung. The importance of megafauna for transferring essential nutrients was highlighted in a study that found the removal of 70 percent of animals weighing over 10 kg during extinctions in the Amazon reduced the lateral movement of phosphorus by more than 98 percent.[161]

In South Africa's Kruger National Park, megafauna like wildebeest are essential for the lateral transport of limiting nutrients like phosphorus from areas of excess nutrients to areas in deficit. Through this process, they increase productivity in the whole park. Because megafaunal species were once more abundant, researchers studying the ecosystem services at Kruger report that some of today's other nutrient-poor regions, like the Amazon Basin, the Congo Basin, and Southeast Asia, may have had much more bioavailable phosphorus in the distant past, when megafauna were more abundant even without the mineral-releasing effect of glaciation. Livestock managed by people could theoretically perform this previously unrecognized nutrient translocation service in the absence of sufficient numbers of large, wild migrating animals.[162] This is exactly the kind of tool that can benefit efforts to rewild the planet if the animals are managed holistically.

## Fertilizer Was Flown onto the Land

In ancient Australia, seabirds would also have facilitated the transfer of essential nutrients, including nitrogen, phosphorus, zinc, and potassium. Seabirds ate fish, absorbing nutrients from the ocean and transferred them to shore in their feathers and dung (or guano). This fertilized the coastal estuaries, grasslands, and forests where they roosted and nested. Today, seabirds transport as much as 3.8 million tons of nitrogen and 630,000 tonnes of phosphorus from the sea to islands where they nest.[163] The power of this important "circular seabird economy" is just a mere remnant of what it was before populations of seabirds were diminished by people through bycatch during fishing, the introduction of invasive predators, and plastic pollution. Today the future of the circular seabird economy is clouded by the fact one-third of seabird species are near extinction.[164]

**161** Doughty et al., "The Legacy of the Pleistocene Megafauna Extinctions," 1.

**162** Wolf et al., "Lateral Diffusion of Nutrients," 9.

**163** Holly P. Jones et al., "The Circular Seabird Economy Is Critical for Oceans, Islands and People," *Nature Reviews Biodiversity* 1, (2025): 691, https://doi.org/10.1038/s44358-025-00099-w.

**164** Jones et al., "The Circular Seabird Economy," 697.

It is hard for contemporary people to believe that bird guano could make a significant difference in the productivity of ancient ecosystems, but if Australian bird flocks were anywhere as abundant as birds in pristine North America, even up to the 1800s, the idea is plausible. In 1813, naturalist and wildlife painter John James Audubon reported a single migrating flock of native passenger pigeons in Kentucky that passed overhead, undiminished, for three days! Scientists have estimated the size of this flock to be 2.2 billion birds.[165] Audubon testified, "The air was literally filled with pigeons; the light of noonday was obscured as by an eclipse; the dung fell in spots, not unlike melting flakes of snow."[166] This would equate to an application of foliar fertilizer, as plants can take up liquid nutrients through their leaves, not just their roots. The guano of American passenger pigeons piled up beneath roosts and nesting colonies that covered thousands of square kilometers, showcasing an extraordinary ability to transport and deposit nutrients across large distances.

This account of fantastic nutrient cycling and productivity in pristine environments has mostly been erased from the modern mind after the very last of three billion passenger pigeons succumbed to guns and deforestation in just 50 years.[167] There is every reason to believe that birds were once abundant across most regions of the world, including Australia, before human settlement. The insects that sustained the birds could have transported twice at many nutrients.

## The High Heel Test

The 2-m-tall giant flightless bird *Genyornis newtoni* had feet that resembled hooves. Why the adaptation of hoof-like feet on a bird? Sounds strange, but it makes perfect sense to me. They help to fertilize the soil.

For years, I wondered why my cows and other ungulates evolved with hard hooves that were so slippery on winter ice. Aurochs, the giant ancestors of modern cows, evolved in Eurasia, which has harsh winters like those in New York and Vermont where I have farmed. Yet, sometimes in winter if it was icy, my cows couldn't walk outside without slipping and falling; it was like they were wearing stiletto heels on a skating rink! In winter, I asked my employees to walk outside and imagine they were wearing high heels instead of insulated work boots to decide if it was safe for cows to walk outside without slipping and falling. I dubbed this the "high heel test."

**165** John H. Schulz et al., "100th Anniversary of the Passenger Pigeon Extinction: Lessons for a Complex and Uncertain Future," *Wildlife Society Bulletin* 38, no. 3 (2014): 445, https://doi.org/10.1002/wsb.453.

**166** Audubon, as quoted in Harold Jones, *Audubon Gallery Catalog* (Manlius Library, 2017): 40.

**167** Schulz et al., "100th Anniversary of the Passenger Pigeon Extinction," 446.

Could hard hooves for fighting and fleeing predators have made ungulates vulnerable in winter compared to lions and bears with better-gripping, padded feet armed with claws? Maybe, but the benefit of hard hooves for fertilizing the grassland feeding them outweighed any winter disadvantage. It turns out hooves are not just for protection. They help grazers keep grasslands healthy.

Stirton's thunderbirds (*D. stirtoni*), and other birds in the *Dromornis* genus, evolved as integral elements of the Australian grasslands. Hoof-like feet may have helped these birds cycle nutrients on grasslands faster by hammering uneaten dead grass leaves and stems down to the ground, where termites, fungi, and other soil biology could use them for food on nature's conveyor belt of nutrient cycling. Cows and wild hoofed animals can do this today if they are managed to mimic their ancestors. By turbocharging nutrient cycling, plants grow faster, allowing even more animals to graze. Birds and other wildlife would also have more food to eat. The superpower megafauna possess to jump-start the productivity of the entire ecosystem helps explain the unexpected abundance seen in fossils discovered near Kati Thanda-Lake Eyre.

## Biology and Climate Are Linked by Water

Climate and biology are not separate now and never were, although they are separate areas of scientific study. Just like the Amazon rainforest creates its own rain,[168] Kati Thanda-Lake Eyre and many other lakes and chains of ponds dotting the now-arid interior of Australia were a result of the rain-producing effect of green plants and trees. These in turn were kept healthy by the unique fauna that had coevolved with them over 20 million years. Regular rainfall and a properly functioning water cycle were the lever connecting biology to the atmosphere that was required to create the megalakes and other large lakes, hydrating the continent. Evaporation would have needed to be one fifth of what it is today to keep Kati Thanda – Lake Ayre full.[169] This required more clouds. A protective green cover of trees, shrubs, grasses, fungi, and herbs supporting life above and below the soil surface provided those conditions.

**168** Jonathon S. Wright et al., "Rainforest-Initiated Wet Season Onset over the Southern Amazon," *Proceedings of the National Academy of Sciences of the United States of America* 114, no. 32 (2017): 8481, https://doi.org/10.1073/pnas.1621516114.

**169** Webb, "Palaeotrophic Reconstruction," 321.

## How Humans Caused the Extinction of Australian Megafauna

When the first Australian people arrived, they encountered naive creatures that were not adapted to flee from human predators. The giant short-faced kangaroos (*Procoptodon goliah*) were more than twice the weight of red kangaroos (*Osphranter rufus*), the largest modern kangaroo. Due to their bulk, and with feet that were equipped with one hoof-like toe, they may have walked instead of hopping. Unlike smaller, more agile kangaroos, these marsupial giants had size and strength that was enough of a deterrent to keep them from being wiped out by marsupial lions at the time. But they were probably vulnerable to human hunters, who trapped them or threw spears at them from a distance using woomeras.

Giant flightless birds may have been extremely susceptible to human predation. Using ancient protein analysis, scientists have confirmed that a burnt eggshell dated to 50,000 years ago belonged to the giant, flightless bird *Genyornis newtoni*, revealing that some first Australian peoples may have eaten their eggs.[170] This kind of predation would have been something these flightless birds had never been exposed to in their millions of years of evolution. This made them vulnerable to rapid exploitation like that which occurred upon human arrival in New Zealand.[171]

## Breaking Australian Mineral Cycles

Human-initiated landscape changes from hunting were profound on the unique but fragile Australian continent.[172, 173] I argue that this set off a cascade of ecological disruption that resulted in desertification. Combined with human-initiated ecosystem decline in southeast Asia and elsewhere at the time, this marked the first known instance of human impact on the global climate. Similarly, the key to a cooler, greener climate lies within this human ability to change climate through land management changes on vast areas of land.

**170** Beatrice Demarchi et al., "Ancient Proteins Resolve Controversy over the Identity of *Genyornis* Eggshell," *Proceedings of the National Academy of Sciences of the United States of America* 119, no. 43 (2022): 6, https://doi.org/10.1073/ pnas.2109326119.

**171** Richard N. Holdaway et al., "An Extremely Low-Density Human Population Exterminated New Zealand Moa," *Nature Communications* 5, no. 1 (2014): 6, https://doi.org/10.1038/ncomms6436.

**172** Miller et al., "Ecosystem Collapse," 290.

**173** Susan Rule et al., "The Aftermath of Megafaunal Extinction: Ecosystem Transformation in Pleistocene Australia," *Science* 335, no. 6075 (2012): 1483–86, https://doi.org/10.1126/science.1214261.

Madjedbebe, a rock shelter in Australia's far north, provides the earliest known evidence of human occupation on the continent, dating back between 60,000 and 70,00 years.[174] The first settlers were probably a relatively small population compared to the groups who arrived at Paisley, at the southernmost part of Western Australia, around 47,000 years ago. By then, human populations had grown large enough to wipe out the megafauna near Paisley within 4,000 years,[175] implicating humans as the drivers of megafaunal extinction in Australia.

Some First Peoples in Australia use boomerangs, spears, and nets to hunt birds today. Could their ancestors have also hunted bird chicks, like the Māori in New Zealand who prey upon tītī, or muttonbird chicks, today? Could fewer birds have short-circuited the nitrogen and phosphorus cycle of the Australian continent?

Modern First Nations Barngarla people trap fish on the Eyre Peninsula in South Australia using vast fish traps built from stone, and according to oral history, there were once many more.[176] Fish and insects also transfer nutrients from the seas and estuaries inland. In Australia, fish traveled up rivers to connected chains of billabongs, or ponds, for millions of years. There, they and their eggs were eaten by large fish, crocodiles, and large species of platypus, successfully transferring the nutrients from the sea to terrestrial predators and regions farther inland. Migrating salmon do this today in North America, where constructed dams do not hinder their migration. In recent decades, salmon have brought an average of 9,000 tonnes of nitrogen and 1,100 tonnes of phosphorus every year to inland ecosystems from the Pacific Ocean.[177]

On the rich landscapes of Australia's lost past, many species of animals helped transport the fertility brought onshore by seabirds farther inland. The marsupial *Diprotodon*, many species of kangaroo, eleven species of wallaby (*Protemnodon*), emu (*Dromaius novaehollandiae*), and other megafauna grazed areas that had been fertilized by seabirds. As animals moved about over millions of years, they ingested the sea-sourced nutrients that plants had taken up and spread them gradually inland in the form of dung and dead carcasses. Forest and grassland-dwelling bird flocks, including 500-kg and 3-m-tall Stirton's thunderbird (*Dromornis stirtoni),* 250-kg "demon duck of doom" (*Dromornis planei*), the largest

**174** Chris Clarkson et al., "Human Occupation of Northern Australia by 65,000 Years Ago," *Nature* 547, no. 7663 (2017): 309, https://doi.org/10.1038/nature22968.

**175** Van der Kaars et al., "Humans Rather than Climate," 4–5.

**176** Ghil'ad Zuckermann and Emmalene Richards, *Mangiri Yarda: "Healthy Country": Barngarla Wellbeing and Nature,* (Revivalistics Press, 2021). 25–26.

**177** Jessica E. Brandt et al., "Continental-Scale Nutrient and Contaminant Delivery by Pacific Salmon," Nature 634, no. 8035 (2024): 876, https://doi.org/10.1038/s41586-024-07980-2.

Australian bird of prey (*Dynatoaetus gaffae*), and a plethora of other bird species shifted nutrients laterally too as they went around in search of food or migrated. This amount of nutrient transfer would not have been trivial. Any reduction in the abundance of animals and insects would have put the brakes on nutrient cycles that were so critical for plant growth on Australia's ancient, mineral-depleted soils.

Not surprisingly, hunting megafauna to extinction in ancient times reduced the productivity of entire ecosystems, just as removing animals from landscapes does today. For instance, scientists in Gabon found that the hunting of forest elephants and other megafauna reduced carbon cycling by 25 percent. When elephant numbers were reduced, termite populations dramatically declined too.[178] The carbon-rich trees and brush that the termites relied upon for a food source were no longer being trampled to the soil surface. Termites play an important role in tropical African forests by cleaning up brown, trampled plants and elephant dung. They farm fungi in expansive nests to transform this carbon-rich biomass into energy-dense food for animals farther up the food chain—a very sophisticated upcycling of carbon in a circular carbon economy. The elimination of many species of megafauna in Australia would have severed countless symbiotic relationships with other life that we will never know about and pushed entire ecosystems past tipping points.

## Short-circuiting the Water Cycle

The community of life that self-organized eons ago in Australia formed a green, unfamiliar world that depended upon its megafaunal gardeners for its existence. Megafaunal extinctions did not happen overnight. But, thousands of years after the first human hunters arrived, many giant animal species had been wiped off the planet. Could ecosystem decline due to human hunting have accelerated the aridification of Australia, just as human impact had hastened the drying of the Sahara?[179]

Scientists previously thought that the African monsoon responsible for intermittent green periods in the Sahara over the past 200,000 years was triggered solely by the Earth's orbital changes. But research in 2020 found that the influence of green plants, the presence of lakes, and the quality of the soil were all essential

---

**178** Therese Lamperty et al., "Defaunation of Large Mammals Alters Understory Vegetation and Functional Importance of Invertebrates in an Afrotropical Forest," *Biological Conservation* 241, 108329 (2020): sect. 3.1, https://doi.org/10.1016/j. biocon.2019.108329.

**179** Wright, "Humans as Agents in the Termination of the African Humid Period."

elements in ensuring the green landscape persisted for as long as it did.[180] Could this also be true for the Australian monsoon? Did climate drive biology, and did biology drive climate? Which came first, the chicken or the egg? Large marsupials and reptiles nurtured the green plants that protected the soil, facilitating both water infiltration and transpiration, which helped in the formation of rain. This was critical in preventing Central Australia from drying out.

The last time Australian megalakes were full was about 48,000 years ago when Kati Thanda-Lake Eyre was 20 m deep. This coincides with the time many megafauna started to go extinct,[181] and the region saw a climatic shift from more stable lake levels to what we see today, with long periods of drought punctuated by episodic flooding.[182] Could human hunting of megafauna have removed the symbionts essential for the high ecosystem productivity required for water cycles healthy enough to keep mega lakes full? Did human-ignited fire and ineffective trampling of old plants by fewer and smaller remaining grazing species initiate desertification? Climate change alone is unlikely to have been the primary cause of the aridification of the Kati Thanda-Lake Eyre basin, because the climate changes that occurred between 60,000 and 45,000 years ago were far less severe than previous climate fluctuations the ecosystem had endured.[183]

At Lynch's Crater, Queensland, in the humid northeast, a decline in the spores of dung fungi between 42,000 and 40,000 years ago shows that a drop in the abundance of grazing animals coincided with human arrival to that region. This occurred just before the forests experienced a dramatic ecosystem shift, so it was unlikely that climate change was to blame for the sudden decline in animal abundance. The forest makeup then changed from a diverse mix of rainforest trees and drought-adapted and fire-adapted trees (sclerophyll), to forests containing only fire- and drought-adapted trees, including eucalyptus.[184]

The decline of the Australian rainforest in its northern reaches may have been the result of intensive hunting of animals and birds that had previously maintained tree species diversity by spreading tree seeds. Cassowary, a large flightless native bird in Queensland today, is responsible for spreading some large tree seeds that no other animals can manage to eat whole. The seeds of some trees will not germinate unless they pass through the digestive tract of the now-rare cassowary. There were likely countless similar symbiotic partnerships

**180** Chandan and Peltier, "African Humid Period Precipitation," sect. 4, para. 1.

**181** Miller et al., "Ecosystem Collapse," 287.

**182** Cohen et al., "Hydrological Transformation," 2.

**183** Miller et al, "Ecosystem Collapse," 290.

**184** Rule et al., "The Aftermath of Megafaunal Extinction," 1486.

with plants and animals that allowed rainforests in Australia, like rainforests around the world, to self-organize into high states of productivity thanks to the power of diversity. Rainforest trees transpire vast quantities of water into the atmosphere to form clouds. The "biotic pump" process may have drawn moisture inland from the sea even in winter, when the Australian summer monsoon was not active. Like the Amazon forest today, Australian forests may have retranspired rainwater several times to create a domino effect of rain, pushing it farther inland.[185]

## Australia Is Stripped of Her Green Cloak

After Australia's first exposure to hunting and fire-wielding people, lower productivity in less species-rich rainforests would have meant less rain to maintain the forest and grasslands that fed the inland lakes. Less water meant less photosynthesis to feed the fauna, one of many negative feedback effects. Thus, the initiation of human hunting on the continent would have been a shock to the productivity of the fragile savanna and rainforest ecosystems, possibly even pushing them past tipping points to new, less productive regimes within several thousand years of human arrival.

A team of scientists documented just such a severe ecosystem shift in Central Australia. By studying $^{13}C$:$^{12}C$ ratios in eggshells and teeth of extinct megafauna, they found solid evidence for an abrupt ecological shift when humans arrived and megafaunal extinctions peaked about 50,000 to 40,000 years ago. They suggest that the use of fire on the landscape by humans may have degraded the drought-adapted savanna landscape, composed of a diverse mosaic of trees, shrubs, and nutritious grassland. The fire-adapted desert scrub that replaced the once species-rich landscape could not support all the animal species that required a wide complement of productive plants, causing the extinction of many.[186] But the removal of many large grazing and browsing animals alone could have played a dominant role in the demise of the unlimited rainforest, open woodlands and savannas.

Human hunting of many large grazing animals that ate flammable grass and brush could have resulted in more wildfire and shifted the ecosystem to the more impoverished landscape of fire-adapted trees and grasses that have taken over most Australian landscapes ever since as occurred at Lynch's Crater.[187]

**185** Van der Ent et al., "Origin and Fate of Atmospheric Moisture over Continents," sect. 1.

**186** Miller et al., "Ecosystem Collapse," 290.

**187** Rule et al., "The Aftermath of Megafaunal Extinction," 1485-1486.

Wild megafauna previously recycled the carbon in dead plant material and controlled brush through their presence. The setting of small cool-burning fires to control woody fuels, aka "firestick farming," may have been an adaptation by the first Australians and other first peoples to replace this lost ecosystem service. Landscapes around the globe have been fundamentally changed by the indirect and direct effects of early fire-wielding humans.

Scientists working in Malawi have identified the earliest known evidence of a link between human behavior and ecosystem transformation with increased fire use dating as far as 85,000 years ago. Fire became a destructive tool that reduced floral biodiversity. Ancient sediments revealed that significant soil erosion had occurred in Malawi, probably due to the burning of forests by humans.[188] The use of fire as a tool is still destructive today when it is used to replace the carbon-recycling activities of megafauna and clear brush or grass stubble. The carbon held in plant material then is released into the atmosphere as a pollutant instead of becoming food for fungi and microbes and building soil carbon.

By 40,000 years ago, Australia had lost 88 percent of its megafaunal species.[189] As the Kati Thanda-Lake Eyre catchment started to desertify as a result, the water cycle would have been damaged. After the soil carbon sponge on this immense area oxidized into the atmosphere, rain would no longer soak into soil and reemerge slowly to feed perennial springs.

Large perennial river inflows into Kati Thanda-Lake Eyre were required over centuries to maintain the much higher water levels seen prior to 70,000 years ago. The bigger the lakes got, the more water was required to offset evaporation. Highly-functioning productive savanna and forests would have provided these conditions. But major hydroclimatic shifts occurred between 70,000 and 60,000 years ago, when river flows became much lower in the Lake Eyre catchment (**Figure 7.1**). Could this be the smoking gun that signals human impact on the paleoclimatic record?

Not enough scientific attention has been given to the possibility that early human impact on ecosystems played a role in the hydrological changes that occurred between 75,000 and 60,000 years ago in Australia because many

**188** Jessica C. Thompson et al., "Early Human Impacts and Ecosystem Reorganization in Southern-Central Africa," *Science Advances* 7, no. 19 (2021): 6, https://doi.org/10.1126/sciadv.abf9776.

**189** Anthony D. Barnosky, "Megafauna Biomass Tradeoff as a Driver of Quaternary and Future Extinctions," *Proceedings of the National Academy of Sciences of the United States of America* 105, (2008): 11543, https://doi.org/10.1073/ pnas.0801918105.

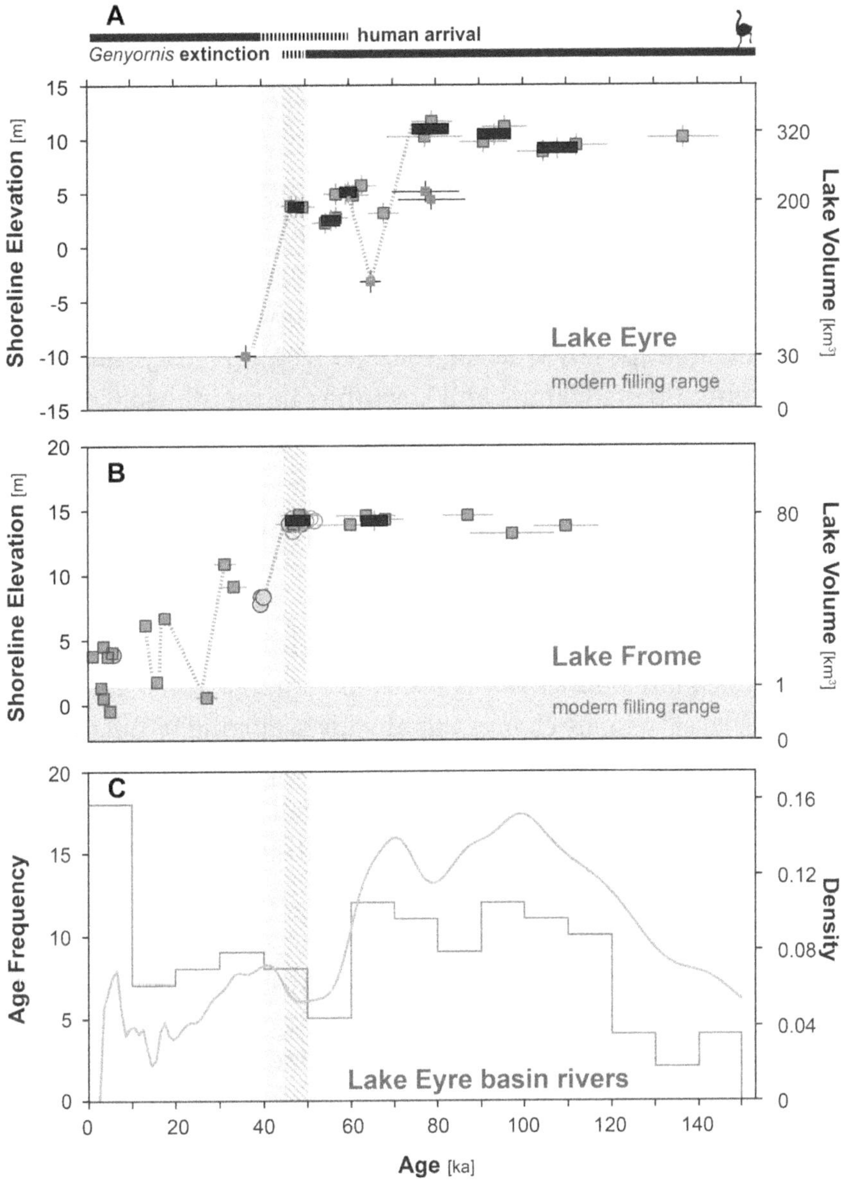


**Figure 7.1 Timelines for human arrival on Australia and continent-wide megafaunal extinction (vertical gray band) including *Genyornis* (hashed grey band) A: Lake Eyre shoreline ages. B: Lake Frome shoreline ages. C: River activity in the Lake Eyre basin.**

Source: Cohen et al., 2015.

scientists don't think people had arrived yet. I think this geoclimatic evidence is proof that they had, supporting the work of Clarkson et al., in 2017, that put human arrival in Australia at about 65,000 years ago.[190]

190 Chris Clarkson et al., "Human Occupation of Northern Australia by 65,000 Years Ago," *Nature* 547, no. 7663 (2017): 306, https://doi.org/10.1038/nature22968.

The decline of productive savanna ecosystems may be the reason the Kati Thanda-Lake Eyre catchment became drastically drier between 65,000 and 45,000 years ago. Removing ecosystem engineers, such as *Obdurodon*, (a platypus-like animal, but twice as large) which wallowed in the many lakes and ponds, could have also changed landscapes significantly, like removing beavers does today. In Canada and the US, beavers act as engineers of riparian areas that increase diversity and abundance of species in their ecosystem.[191] Large wading and swimming Australian marsupials and reptiles slowed the movement of water across ancient landscapes by creating soaks and sumps to hydrate their green world. The drying up of swamps and savannas in their absence would have released huge amounts of carbon into the air.

## The Golden Spike Marking the Start of the Anthropocene

Kati Thanda-Lake Eyre was already starting to dry up when people finished exterminating the giant bird *Genyornis newtoni* between 50,000 and 45,000 years ago, and some suggest climate change was already occurring.[192] But these changes to regional climate could have been initiated by humans who arrived in Australia 20,000 years earlier. Ice core evidence reveals that a global temperature and atmospheric $CO_2$ rise occurred between 63,500 and 60,000 years ago.[193] Scientists consider this global event particularly unusual because the $^{13}C$ to $CO_2$ ratio ($^{13}C$-$CO_2$) in sediment records dropped at an unprecedented rate. This suggests that the addition of $CO_2$ into the atmosphere at the time was originally sourced from living organisms and not from volcanic activity or the geochemical process of $CO_2$ dissolving more slowly in a warming ocean.[194] Could this represent a golden spike in the geological record, marking an early start of the Anthropocene?

This aberration in the $^{13}C$ ratio to $CO_2$ between 63,500 and 60,000 years ago was the most pronounced in 140,000 years.[195] It was unmatched until the Industrial Revolution, when fossil-fuel combustion released massive amounts of ancient, previously sequestered carbon (originally taken out of the atmosphere by photosynthesis), back into the atmosphere (**Figure 7.2**).

---

**191** Emily Fairfax and Cherie Westbrook, "The Ecology and Evolution of Beavers: Ecosystem Engineers that Ameliorate Climate Change," *Annual Review Ecology, Evolution, and Systematics*, 55, (2024): 337, https://doi.org/10.1146/annurev-ecol-sys-102722-122317.

**192** Cohen et al., "Hydrological Transformation," 195.

**193** James A. Menking et al., "Multiple Carbon Cycle Mechanisms Associated with the Glaciation of Marine Isotope Stage 4," *Nature Communications* 13, no. 1 (2022): 2, fig 1, https://doi.org/10.1038/s41467-022-33166-3.

**194** Menking et al., "Multiple Carbon Cycle Mechanisms Associated with the Glaciation of Marine Isotope Stage 4," 1–2.

**195** Menking et al., "Multiple Carbon Cycle Mechanisms," 2.

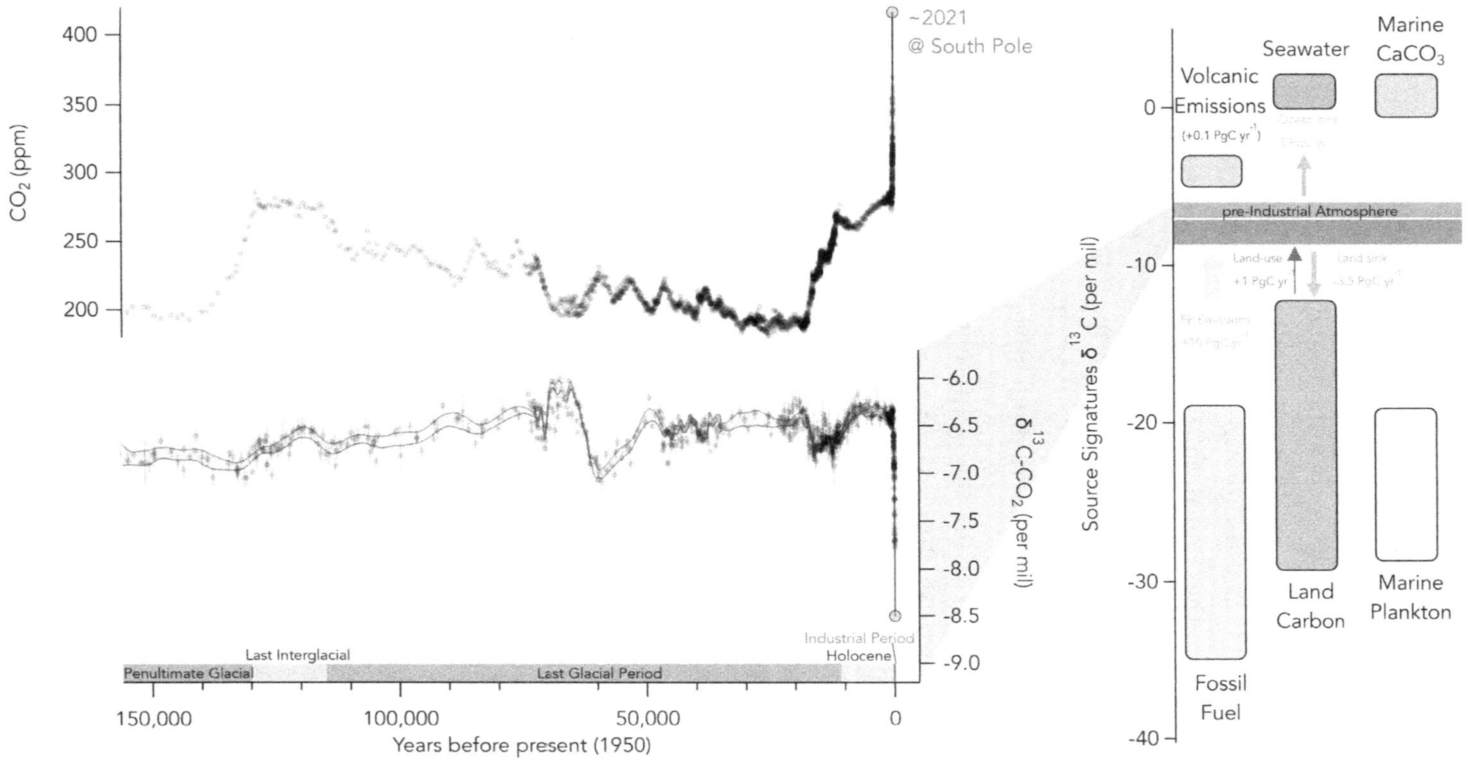


Data sources: Ice Cores - Marcott et al., 2014; Bereieter et al., 2015; Bauska et al., 2015; Bauska et al., 2016; Eggleston et al., 2016; Bauska et al., 2018; Rubino et al., 2019; Bauska et al., 2021; Menking et al., 2022.
Instrumental Records - South Pole Station data from scrippsco2.ucsd.edu (Keeling et al., 2001). Current carbon budget (simplified) - https://www.globalcarbonproject.org. Draft Figure - Thomas Bauska (thausk@bas.ac.uk)

**Figure 7.2 LEFT PANEL: The concentration and isotopic composition of atmospheric $CO_2$ from near the present day to the second to last glacial period. RIGHT PANEL: a plot showing an approximate isotopic range of the major sources of $CO_2$.**

Reproduced from Thomas K. Bauska, 2025

The carbon cycle is complex and no one factor can explain the $^{13}C-CO_2$ anomaly, but a rapid influx of land carbon into the atmosphere could have contributed to the abrupt change in the $^{13}C-CO_2$ ratio. Scientists who looked at the abundance of dung spores at Caledonia Fen, a site in southeast Australia, noticed a sharp drop in dung spores associated with megafauna like *Diprotodon* between 76,000 and 60,000 years ago, soon after humans arrived in Australia.[196] The removal of many megafauna could have pushed the diverse, hydrating ecosystems that had evolved over millions of years past tipping points to new fire-adapted but less productive regimes. The fact that scientists document historically unusual low water levels at Kati Thanda-Lake Eyre beginning then supports the hypothesis that the source of extra atmospheric $CO_2$ at the time came from terrestrial sources, as immense areas like the Kati Thanda-Lake Eyre catchment lost soil carbon as it became less productive.

Australia was not the only place being impacted by humans. Human-driven ecosystem productivity decline spread across the Indian Subcontinent and Southeast Asia simultaneously as humans arrived. There was a human population explosion between 75,000 and 33,000 years ago as humans left Africa and encountered new abundant food resources.[197] This would have exacerbated the human impacts of hunting and fire.

Ice core records show that atmospheric $CO_2$ and global temperatures have fluctuated almost in lockstep over the last 800,000 years. It is therefore likely that the warming of air and ocean temperatures seen between 63,000 and 60,000 years ago was influenced by the greenhouse effect of increased $CO_2$. Carbon sequestration via the biological pump in oceans would also have slowed because warmer waters are less biologically productive than colder seas, preferentially slowing light carbon ($^{12}C$) sequestration as well and lowering the $^{13}C-CO_2$ ratio even further. This anomaly in the $^{13}C-CO_2$ in atmospheric $CO_2$ molecules may mark the earliest geologically documented signal of human impact on the global climate (**Figure 7.3**).

**196** Chris N. Johnson et al., "Geographic Variation in the Ecological Effects of Extinction of Australia's Pleistocene Megafauna," *Ecography* 39, no. 2 (2016): 112, https://doi.org/10.1111/ecog.01612.

**197** Alan R. Rogers, "Genetic Evidence for a Pleistocene Population Explosion," *Evolution* 49, no. 4 (1995): 614, https://doi. org/10.1111/j.1558-5646.1995.tb02297.x.

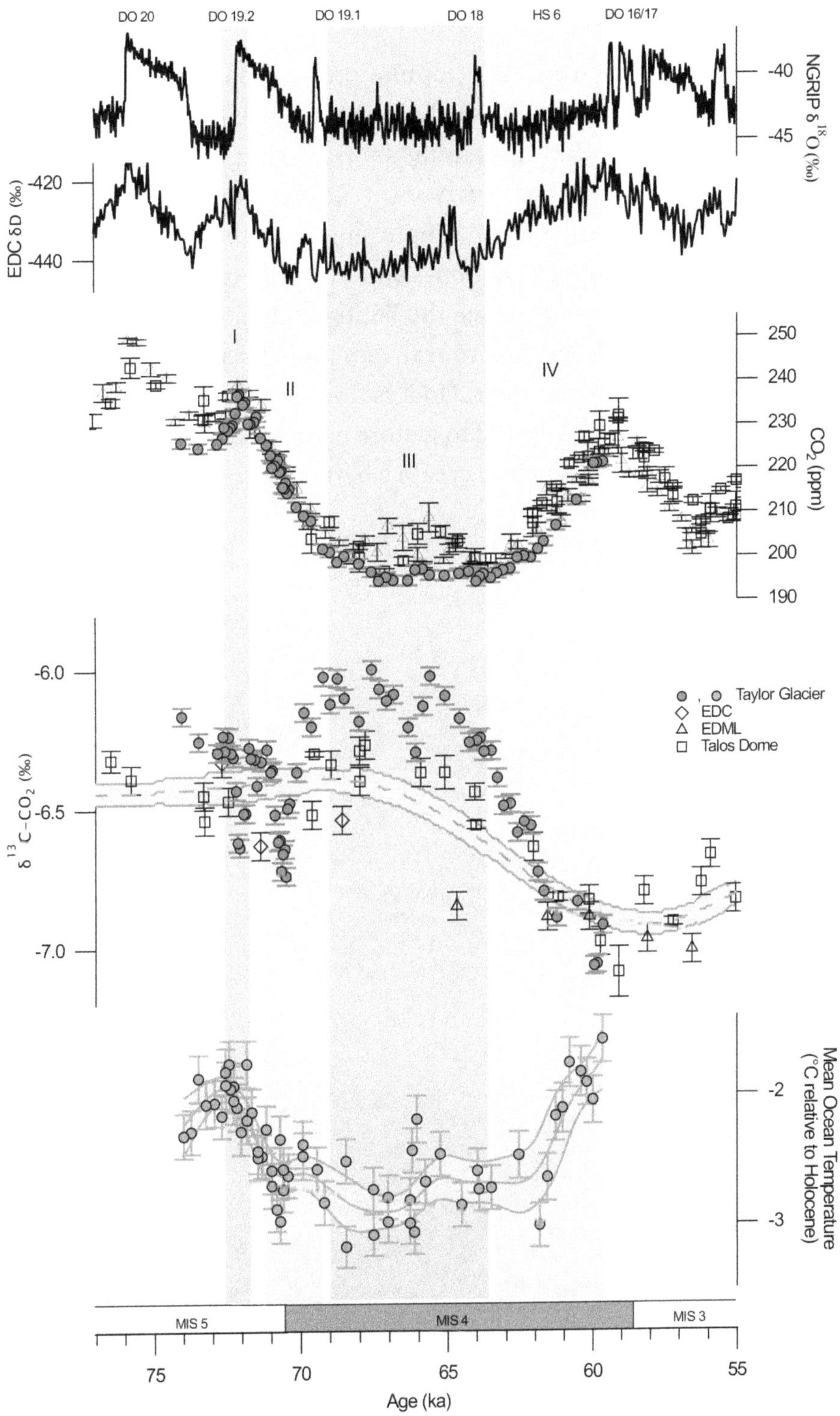


**Figure 7.3 $^{13}C$-$CO_2$, $CO_2$, Ocean temperature from Taylor Glacier.**

Reproduced from Menking, 2022.

## The Story of Hope

If human activity at much smaller population sizes can degrade ecosystems enough to impact global climate, we hold much more power today to restore the Earth's climate-stabilizing systems by repairing damaged ecosystems. This story of hope is what the return of the Karner blue butterfly, bluebirds, and bobolinks to Raindance Farm taught me. To truly believe in ecosystem restoration, you need to see it in action—or better still, take part in it where you live. It makes me so optimistic to see the Youngwilders organization in the UK who are organizing youth-led nature recovery projects across the country, and they are looking for more volunteers. Understanding your land's original healthy state and how animals are essential to restore pristine levels of productivity and diversity can help you envision and create a better future as you plant your own "Garden of Eden."

# Chapter 8

# Our Fall from Grace

Across the globe, savannas and grasslands emerged as recurring and highly productive ecosystem templates, generating vast natural capital in the form of carbon-rich soils. These fertile landscapes became the foundation of modern agriculture, seen today in regions such as the American Corn Belt, the Canadian Wheat Belt, and the chernozem soils of the Ukrainian steppe (known as the Breadbasket of Europe).

With the rise of agriculture and pastoralism around 8,000 years ago, humans began to mine soil with unsustainable farming. We have struggled to replace the soil carbon that we have allowed to shape-shift into $CO_2$ and escape into the atmosphere by tilling our fields. Technology cannot create soil; only living systems can, through the complex processes of fully functioning ecosystems.

## The First Solar Energy Storage

For millennia, people have quite literally "lived off the fat of the land." The ability of nature to capture the sun's energy and produce dense, energy-rich forms of chemical energy to sustain people and other animals is amazing. One gram of fat has twice the energy of a gram of carbohydrates or protein. Gaining a deeper understanding of how optimizing energy flow can help us to restore shrinking natural resources like fertile soil and fresh water is an important step in ensuring the survival of our species.

Mammals are all attracted to the high-energy portions of our food. Cows and other grazers and browsers prefer to eat the tops of plants or the tips of shoots where sugars and lipids are concentrated. Their bodies will transform tough carbohydrates, which are inedible for people, into lactose, protein, and fat. During times of surplus plant growth, wild herbivores must accumulate and store fat in their bodies to draw upon for energy during lean times. The fatty humps on the backs of camels and zebu cattle, and on the rumps and tails of fat-tailed sheep, act as natural solar-energy storage.

Predators also selectively eat the energy-rich portions of their food first as part of their evolutionary strategy. Orcas in New Zealand prey on stingrays and sharks and usually only extract the fat-rich liver, leaving the rest for scavenging ecosystem partners. Even my cat used to start her mouse meals by eating the head of her prey because the brain contains a higher proportion of fat. Sometimes that is all she ate if an energy snack was all she was after. Many grizzly bears plump up on fatty salmon in the autumn to get through their cold winter hibernation. At least 4,000 years ago, humans developed solar-energy storage technology when they learned to store animal fat and protein in the form of cheese, storing summer solar energy for use in winter.

Ancient humans improved their chances of survival by acquiring high-fat foods, which provided more energy and nutrients than lean meat alone. Fossil evidence from a 10,500-year-old buffalo kill site in Cooper, Oklahoma, suggests that Paleolithic humans butchered just the fat-rich hump and rib meat of the animals.[198] After trapping up to 30 bison at a time at the site of a dry stream bed, the hunters dispatched them with stone spearheads, called Folsom points, launched with atlatls. Interestingly, these hunters left the leaner meat of the legs and shoulders behind, showing that they placed more value on fatty hump and rib meat. Perhaps their dogs fed on the meat before wild scavengers cleaned it up.

For Paleolithic people, obtaining energy from animal sources may have been not only essential for survival, but also beneficial for their health. Analysis of 229 modern hunter-gatherer populations whose diets were made up of only 26–35 percent vegetables, but 66–75 percent animal origin products, found no evidence of cardiovascular disease among them.[199]

## Quality of Life in Prehistoric "Gardens of Eden"

Paleolithic bands of *Homo sapiens* had food delivered to their neighborhoods on the hooves, wings, and fins of the animals they depended on. This ancient DoorDash came without money changing hands and without the use of a drop of fossil fuel. The herbivores, birds, reptiles, and fish that made up human diets did all the seed planting, cultivation, fertilizing, soil conditioning, and harvesting of animal feeds for the people, also for free. Their presence even helped to regulate water cycles, ensuring that the rain continued to fall on the plains of the past

**198** Eileen Johnson and Leland C. Bement, "Bison Butchery at Cooper, a Folsom Site on the Southern Plains," *Journal of Archaeological Science* 36, no. 7 (2009): 1445-1446, https://doi.org/10.1016/j.jas.2009.02.007.

**199** L. Cordain et al., "The Paradoxical Nature of Hunter-Gatherer Diets: Meat-Based, yet Non-Atherogenic," *European Journal of Clinical Nutrition* 56, no. S1 (2002): S46, https://doi.org/10.1038/sj.ejcn.1601353.

without the need for aqueducts or irrigation. Such was the power of productive ecosystems to deliver renewable natural resources like water and soil as a built-in feature of biodiverse landscapes.

Like many modern hunter-gatherer societies,[200] Paleolithic societies were likely highly egalitarian. Their economies certainly were circular. All goods required for life, thriving economies, and culture could be sourced locally. All waste, including humanure, was recycled into the soil, enriching it. As Paleolithic cultures developed, they became more efficient at harvesting the most energy-dense food by cooperating with other people and working with dogs and other animals. But there was a fatal flaw. The way we made decisions in lands of abundance worked well to ensure our immediate success but also sowed the seeds of long-term ecosystem decline.

As populations swelled, humans inevitably started devouring the natural capital that underpinned their economies, eating away at the very foundations: soil, water, and wildlife. It's a pattern we continue to repeat today, eroding the ecological base upon which all human prosperity depends. Modern humans act as ecosystem engineers in new areas, by hunting large animals.[201] Evidence also suggests that humans have influenced plant abundance for millennia. For example, the unexpected abundance of useful palms in the Colombian Amazon, including the fruit-bearing tree *Brosimum lactescens*, can be traced back to the gardening efforts of Paleolithic people who lived there as far back as 12,600 years ago.[202]

## Why Did Human Societies Resist Agriculture?

Our ancestors continued the hunter-gatherer lifestyle for hundreds of thousands of years. Our species has only been farming for the most recent 1 percent of our time on Earth.[203] Why did it take so long for humans to adopt agriculture and develop what we call "civilization"? For all mammals, including humans, energy is essential for survival, along with air and water. Evolution rewards behavior that facilitates acquiring the most amount of energy with the least effort.

---

**200** Jessame Thomson, et al., "' Fiercely Egalitarian ' : Thematic Cross-Cultural Analysis Reveals Regularities in the Maintenance of Egalitarianism Across Four Independent African Hunter-Gatherer Groups," *Cross-Cultural Research* Vol. 59, no. 5 (2025): 535–578, https://doi.org/10.1177/10693971251338210.

**201** Therese Lamperty et al., "Defaunation of Large Mammals Alters Understory Vegetation and Functional Importance of Invertebrates in an Afrotropical Forest," *Biological Conservation* 241 (2020): 108329, sec. 4.2, https://doi.org/10.1016/j. biocon.2019.108329.

**202** Gaspar Morcote-Ríos et al., "Colonisation and Early Peopling of the Colombian Amazon during the Late Pleistocene and the Early Holocene: New Evidence from La Serranía La Lindosa," *Quaternary International* 578 (2021): 15–16, https://doi. org/10.1016/j.quaint.2020.04.026.

**203** Vernon L. Smith, "Hunting and Gathering Economies," *in The New Palgrave Dictionary of Economics* (Palgrave Macmillan UK, 2008): 6039, https://doi.org/10.1057/978-1-349-95121-5_1066-2.

Scientists have proposed that it was simple economics that kept our ancestors hunting and gathering.[204] The economic gain per caloric effort invested in hunting was highest when the largest megafauna species were on the menu. This would explain why humans hunted them first until they became scarce, before they set sights on the next-largest species.[205]

It didn't take much energy for hunters, cooperating with other members of the tribe, to whittle spears and harvest enough food to feed the extended family. Strategic cooperation by hunters to take down large-bodied prey gave a high return on energy invested. If each hunter was responsible for feeding a family of several children, a partner, and one or two sets of grandparents, then bringing home a rabbit wouldn't be enough.

Human hubris has fueled the belief that the switch to farming by our Neolithic ancestors was another step in the evolution of civilization toward greater productivity. However, scientific studies reveal that early farmers were only 60 percent as productive as their foraging contemporaries. Hunter-gatherer societies living in the Great Basin grasslands in what is now the US, between 650 AD and 1650 AD, harvested 2,500 calories (kcal) for every hour of labor spent hunting and gathering. In contrast, early maize cultivation in the US yielded only 1,000 kcal per hour of effort. Subsistence farming communities of the 1950s in Mexico, the Philippines, and Borneo produced less than 1,500 kcal per hour from crops such as corn and rice.[206] These comparisons of hunting versus agriculture suggest that early humans' shift to agriculture wasn't an effort to be more efficient. Rather, because people had hunted the megafauna to extinction and foraged most of the dwindling wild resources, they were left with few alternatives. Agriculture was not a great advance. It was our consolation prize for damaging the ecological richness of our inheritance.

There is shocking evidence that human evolution took a step backwards after the switch to farming. Since the late Stone Age (between 30,000 and 2,000 years ago), human skull capacity and by extension, brain size, had decreased significantly in Africa. Brains shrank by 95–165 $cm^3$ in males (about the volume of a cricket ball)

**204** Vernon L Smith, "Hunting and Gathering Economies," *In The New Palgrave Dictionary of Economics*, 1–7. Palgrave Macmillan UK (2008): 6042–43. https://doi.org/10.1057/978-1-349-95121-5_1066-2.

**205** Jacob Dembitzer et al., "Levantine Overkill: 1.5 Million Years of Hunting down the Body Size Distribution," *Quaternary Science Reviews* 276: 107316. (2022): 7, sec. 4, https://doi.org/10.1016/j.quascirev.2021.107316.

**206** Samuel Bowles, "Cultivation of Cereals by the First Farmers Was Not More Productive than Foraging," *Proceedings of the National Academy of Sciences of the United States of America* 108, no. 12 (2011): 4763, https://doi.org/10.1073/ pnas.1010733108.

and by 74–106 $cm^3$ in females.[207] This trend is consistent with a study done on global populations that observed a correlation between lower meat consumption due to megafaunal decline and shrinking brain size.[208] The rate of this nearly 11 percent decrease in relative brain size peaked after most civilizations were relying on agriculture some 3,000 years ago, prompting some scientists to suspect that new supportive social structures and a reduced need to remember and manage complex environments to find food led to smaller human brains.[209]

Farming likely had less appeal than hunting because of the extra work and drudgery involved. Also, if a drought or flood hit, the people who were hunting and foraging were more readily mobile and could find better hunting grounds, so they endured less risk. I remember growing a crop of sorghum to feed my cattle and watching it get flattened by flooding caused by Hurricane Irene in 2011. Luckily, 95 percent of my farm was in resilient perennial pasture that recovered by itself at no cost, even where it had been underwater. That was the last annual crop I ever grew. Risk is an economic cost to any business and something to which our Paleolithic ancestors were no strangers. But as they unraveled the threads of the tapestry of life that formerly was self-organizing and resilient, they had no choice but to take on more risky agriculture.

## Losing Paradise

David Wright, an archaeologist who studied the formation of the Sahara Desert, observed that upon the arrival of Paleolithic humans to new regions around the globe, a common three-step pattern occurred. It all hinges on the holy grail of ecosystem primary productivity, the central theme of this book. First, hunter-gatherers began to reduce ecosystem productivity by overhunting the populations of wild ecosystem engineers. Megafauna, including giant predators, were the first animals to be decimated even though they acted as keystone species for the entire ecosystems they inhabited. According to Wright, the resulting cascade of damaging effects on ecosystem function caused "transformation of the biotic environment from one that inherently produced resources for humans into one that must be cultivated to yield resources."[210]

**207** M. Henneberg and M. Steyn, "Trends in Cranial Capacity and Cranial Index in Subsaharan Africa During the Holocene," *American Journal of Human Biology* 5, no. 4 (1993): 477, https://doi.org/10.1002/ajhb.1310050411.

**208** A. Wiercinski, "Has the Brain Size Decreased Since the Upper Paleolithic Period?," Bulletins et *Mémoires de La Société d'anthropologie de Paris* 6, no. 4 (1979): 426–7, https://doi.org/10.3406/bmsap.

**209** Jeremy M. DeSilva et al., "When and Why Did Human Brains Decrease in Size? A New Change-Point Analysis and Insights from Brain Evolution in Ants," *Frontiers in Ecology and Evolution* 9 (2021): 7, https://doi.org/10.3389/fevo.2021.742639.

**210** Wright, "Humans as Agents in the Termination of the African Humid Period," 6.

Only *after* the land was denuded of much of its game and could no longer support human populations did people begin farming. Water, the lifeblood of wild rivers and swamps, was diverted to grow crops. Initially, agriculture lifted output from the land as farmers mined biology in the soil built up over eons. But, as fertile soil was depleted when nutrients were removed by harvest or erosion, the total productivity of the supporting ecosystem declined. In the case of semiarid environments like the Sahara, the eventual result was desert.

According to a 2019 report commissioned by the UN, the degradation of natural capital is accelerating on a global scale. Agricultural production has increased 300 percent since 1970, only because of a massive infusion of fossil energy. Meanwhile, land degradation has reduced the productivity of 23 percent of the global land surface. Up to $577 billion in annual global crop production is at risk from pollinator loss as our ecosystems head toward collapse.[211] Ten million hectares of farmland and 36 billion tonnes of topsoil are lost each year.[212] It is in everyone's interest to understand why this is happening and how we can reverse it. If we were to replenish today the vast swaths of deep topsoil that had evolved around the world, it would require the same process that created it in the first place.

Natural capital-producing conditions were found in ecosystems replete with large grazing and browsing animals and their predators that were unconfined and able to perform their superpower of nutrient cycling, helping plants to be more productive. This can be a useful template to emulate as we evolve into the Restoracene and start writing a better future script for our grandchildren.

## Rethinking Civilization

If farming was not an advancement for human economies, but an adaptation to less abundant large game animals, could our earlier hunter-gather societies have been more socially advanced than we had assumed? With the advent of farming, people have had to endure diminished egalitarianism.[213] Some people always manage to avoid the hard labor of farming by using slavery, feudalism,

---

**211** Intergovernmental Science-Policy Platform on Biodiversity and Ecosystem Services (IPBES), *Summary for Policymakers of the Global Assessment Report on Biodiversity and Ecosystem Services of the Intergovernmental Science-Policy Platform on Biodiversity and Ecosystem Services* (2019), 13.

**212** Pasquale Borrelli et al., "An Assessment of the Global Impact of 21st Century Land Use Change on Soil Erosion," *Nature Communications* 8, no. 1 (2017): 2013, https://doi.org/10.1038/s41467-017-02142-7.

**213** Serge Svizzero and Clem Tisdell, "The Persistence of Hunting and Gathering Economies," *Social Evolution & History* 14, no. 2 (2015): 7.

or the cheap paid labor of mercantile capitalism. The assumption that our hunter-gatherer ancestors in pristine ecosystems had less time for art, music, and culture is probably false and another example of modern human hubris. When wild game and edible plants were abundant, early societies may have enjoyed much more free time to be creative than we often acknowledge.

The ritual around the hunt, with its success fueled by social cooperation, generated both cunning and innovation in early humans. A successful hunt likely was followed by a celebration that reinforced social bonds. The excitement paleolithic hunters enjoyed while hunting mastodons in the grasslands of what is now the Colombian Amazon would have been far more electrifying than today's Super Bowl. Early paleolithic hunters recorded the megafauna they hunted and trapped by painting them on cliffs as long as 12,500 years ago, including giant sloths, horses, and camelids at Serranía de la Lindosa, preserving their story in stone for millennia.[214] (**Figure 6.6**)

## How People Acquire Food Affects the Soil

After our ancestors drove megafauna to extinction across the globe, ecosystems began to unravel. The loss of top consumers in ecosystem food chains has knock-on effects on what plants grow and where, and can change the face of landscapes.[215] This can increase the risk of fire and cause ecosystems that were once carbon sinks to become carbon sources, as happened in the Serengeti when wildebeest numbers dropped because of a virus. Scientists reckon the Serengetti needs one million wildebeests to remain a carbon sink because they help increase soil carbon and reduce wildfire.[216]

What did this ecosystem restructuring look like? Smaller game species may have become more numerous, but they could never make up for the lost fertilizing effect of megafauna. Waning productivity weakened the natural water cycle, making landscapes drier. In brittle environments with sporadic rainfall, this and other cascading effects initiated shifts from grassland ecosystems to dry shrubland or deserts. In areas with more consistent rainfall, forests took over the landscape.

**214** Morcote-Ríos et al., "Colonisation and Early Peopling," 16, figure 13.

**215** James A. Estes et al., "Trophic Downgrading of Planet Earth," Science 333, no. 6040 (2011): 303, https://doi.org/10.1126/ science.1205106.

**216** Ricardo M. Holdo et al., "A Disease-Mediated Trophic Cascade in the Serengeti and Its Implications for Ecosystem C," *PLOS Biology* 7, no. 9 (2009): e1000210, Disc., para. 6, https://doi.org/10.1371/journal.pbio.1000210.

But because there were fewer large grazers to control brush and understory growth, fires often became more intense and more frequent.[217, 218, 219, 220, 221]

Instead of addressing the root problem of human populations being out of balance and damaging the ecosystems underpinning our economies, farming allowed humans to bypass nature's energy constraints. A vicious cycle of larger families and tighter child spacing emerged in early farming societies, as women's fertility rates increased compared to their hunter-gatherer counterparts. In nomadic cultures, childbearing was naturally more spaced because women had to carry children who were too young to walk.[222]

Agriculture marked the beginning of a new acceleration of the erosion of the natural capital of the commons. By cultivating land primarily for human use, people have appropriated the resources once shared among countless other species. The result is that humans and their livestock, who make up just 0.01 percent of the planet's species, consume most of the world's biological resources. The total wild mammal biomass on the planet has been reduced to only 20 million tonnes, while our livestock has increased to 630 million tonnes, and humans to about 390 million tonnes.[223] This is not harmony. It is cacophony, as noisy city streets continue to drown out the sound of sweet birdsong.

The unraveling of the Earth's biogeoclimatic systems by humans does not have to be our destiny. We now have powerful tools for understanding and managing complex ecosystems that our Paleolithic ancestors didn't have access to. Approaches like holistic management can help us restore the integrity of natural systems and move toward regeneration and coexistence. If we start to respect and restore the grace of nature and finally reconnect with the living world harmoniously, then humans may enjoy this beautiful blue/green planet for another two million years.

---

**217** Estes et al., "Trophic Downgrading," 304.

**218** Jacqueline Gill et al., "Pleistocene Megafaunal Collapse, Novel Plant Communities, and Enhanced Fire Regimes in North America," *Science* 326, no. 5956 (2009): 1103. https://doi.org/10.1126/science.1179504.1103.

**219** Allison T. Karp et al., "Global Response of Fire Activity to Late Quaternary Grazer Extinctions," *Science* 374, no. 6571 (2021): 1147, https://doi.org/10.1126/science.abj1580.

**220** Perrotti et al., "Diverse Responses of Vegetation and Fire after Pleistocene Megaherbivore Extinction Across the Eastern US." *Quaternary Science Reviews* 294: 107696. (2022): sect. 5, https://doi.org/10.1016/j.quascirev.2022.107696.

**221** Julia Rouet-Leduc et al., "Effects of Large Herbivores on Fire Regimes and Wildfire Mitigation," *Journal of Applied Ecology* 58, no. 12 (2021): 2693, https://doi.org/10.1111/1365-2664.13972.

**222** Richard Borshay Lee, *The !Kung San: Men, Women, and Work in a Foraging Society* (Cambridge University Press, 1979), xxi.

**223** Lior Greenspoon et al., "The Global Biomass of Wild Mammals," *Proceedings of the National Academy of Sciences of the United States of America* 120, no. 10 (2023), fig. 4, https://doi.org/10.1073/pnas.2204892120.

# Chapter 9

# Global Traders Write an Impoverished Future Script

When I was growing up in the 1960s and 1970s, I spent many summers staying with my cousins in Ireland. We had a lot of freedom to run feral and explore our natural environment. We tasted green apples in the farm orchard to see if they were ripe, collected chicken eggs and foraged for wild mushrooms for breakfast, and hiked in the hills. My grandparents lived right across the street and loved to share their wisdom. My grandfather, John Malone, showed my city-raised siblings and me that people can grow most of their vegetables in small backyard plots. There, my brother and I were given the job of picking from the soil the bright new potatoes that Granddad revealed to us from under dark soil with his garden fork.

Sometimes we played around Dolly the donkey as Granddad, a town resident, harnessed her up to take the cart out to the field he rented to grow potatoes and barley on the outside of town. At harvest time, we helped tie bags of barley after they were filled by the combine he hired to harvest his grain crop. My great aunt Julia still lived in the thatched cottage on the farm in Killeshin where my ancestors had lived for hundreds of years. She made the best-tasting butter in the world from her grass-fed cows; it was served on hot scones baked over a peat fire. No one had to tell us where our food came from.

Many children growing up today spend most of their days surrounded by concrete, digital screens, and synthetic materials, and are largely living separately from nature. Sadly, they are often deprived of the valuable immersion in nature that I had. They don't get the chance to develop an intuitive understanding of how natural landscapes function. This instills an artificial, human-centric view of the world. When marketers, policymakers, and consumers don't respect or even recognize landscapes and agroecosystems for what they are—our planetary life support system—our future is not safe. This disconnect allows people to collectively justify self-destructive decisions by pretending we can live without nature. As an unintended consequence of so-called "civilized" lifestyles, human are destroying the wellspring of life that is the foundation of all economies on this beautiful and unique planet as we hurtle through deep, dark, inhospitable space. When did humans separate from nature in our minds, and why? What are the ramifications of human exceptionalism and what have we lost?

## Agricultural Surplus: The Currency of Inequality

As people ran out of wild local food sources and began to produce food from the land by farming,[224] grain surpluses, originally grown to supplement game and survive dry years, allowed communities to remain in one place and for populations to increase. Eventually, surplus grain became a tradeable commodity that people used to acquire resources they lacked. As agriculture and trade facilitated humans to support ever larger communities, growing urbanization disconnected policymakers further from nature. There have been three history-changing unintended consequences of continued human depletion of ecosystems: inequality, shrinking natural resources, and loss of social and resource autonomy, which all started at the dawn of agriculture.

As crop surplus became currency for trade, the first core value of society to fall was equality. In Paleolithic hunter-gatherer societies where people worked together, every family and tribe had an equal chance at hunting when lands were abundant. But with the advent of farming around 8,000 years ago, inevitably, some people were left to do most of the hard physical work. Tilling the soil in the hot sun was not a highly anticipated event that could make anyone a hero like a successful hunt could. The most cunning and powerful of our ancestors soon found ways to avoid this difficult, repetitive, physical labor.

First-world people continue to do this today by expecting lower-paid migrants or people in poorer countries to do the most difficult and repetitive farmwork for them. For instance, I have a good understanding of how repetitive milking cows can be, and I struggled to find American workers willing to milk our cows so I could take my children to sports after school. Picking tomatoes in the hot Florida sun is equally unappealing to American workers. This is why between 2020 and 2022, the United States Department of Agriculture (USDA) reported that over 60 percent of hired crop farmworkers were foreign-born. Forty-two percent were undocumented.[225] Without them, the cost of US-grown vegetables and dairy products on store shelves would be far higher.

**224** David K. Wright, "Humans as Agents in the Termination of the African Humid Period," *Frontiers in Earth Science* 5 (2017): 6, https://doi.org/10.3389/feart.2017.00004.

**225** US Department of Agriculture, Economic Research Service, "Legal Status of Hired Crop Farm Workers, Fiscal 1991–2022," infographic, accessed January 20, 2025. https://www.ers.usda.gov/topics/farm-economy/farm-labor.

## The Rise of Empire

Early urban centers developed around 6,000 years ago in the Fertile Crescent: the Levant, Mesopotamia, and southwest Iran. Throughout the Bronze Age and up to 900 BC (before common era) in the Near East, societies were politically organized regionally as city-states. But during the Age of Empire between 900 BC and 224 AD, large empires emerged in the Fertile Crescent, starting with the Neo-Assyrian Empire. By examining the distribution of large houses compared to small houses, scientists concluded that as empires grew, so did inequality.[226]

As the extent of imperial trade routes like the Silk Road expanded, they became not only a conduit of goods but also opportunities for imperial elites to tax more people. By Roman times, inequitable production models such as slavery were well established for farming.[227] These schemes could be duplicated to extract other resources like minerals, including gold and silver, that were added to the quiver of currency used by the elite to this day. Although they were a minuscule proportion of the population, elite royal families leveraged slavery to amass more wealth for themselves and the traders who enabled them.

As in China during the Han Dynasty, ancient Roman elites who collaborated with the people at the seat of imperial power were financially rewarded at the expense of most farmers and workers. Before the rise of powerful empires, war and political instability had often redistributed wealth. But with long-term imperial dominance, certain royal families were able to hold on to generational wealth, resulting in a direct correlation between imperial stability and growing economic inequality. The extent to which inequality has now ballooned was highlighted in an Oxfam International report in 2024. It revealed that between 2020 and 2023, $26 trillion (63 percent) of all new wealth had been captured by the richest one percent, while $16 trillion (37 percent) went to the rest of the world put together.[228]

In modern US agriculture, well-connected traders have gamed the system so there are winners and losers in the race to overproduce just a few subsidized commodity crops like corn, soy, and cotton. Family farmers became the clear losers as the number of farms plummeted. Since the 1950s, American farmers have

226 Andrea Squitieri and Mark Altaweel, "Empires and the Acceleration of Wealth Inequality in the Pre-Islamic Near East: An Archaeological Approach," *Archaeological and Anthropological Sciences* 14, no. 10 (2022): 190, https://doi.org/10.1007/ s12520-022-01659-3.

227 Edward E. Cohen, "Inequality," in *Roman Inequality* (Oxford University Press, 2023), 26–27, https://doi.org/10.1093/oso/9780197687345.003.0002.

228 Martin-Brehm Christensen et al., "Survival of the Richest: How We Must Tax the Super-Rich Now to Fight Inequality," Oxfam briefing paper, January 16, 2023, https://policy-practice.oxfam.org/resources/survival-of-the-richest-how-we-must-tax-the-super-rich-now-to-fight-inequality-621477/.

been convinced to grow more and "feed the world" by buying more costly inputs. The resulting overproduction of a handful of commodity crops resulted in lower prices. Millions of farms went out of business as farm net profitability shrank. The USDA reported six million farms in 1940 at the start of the Green Revolution in the US.[229] By the 2022 census, four million were gone.[230] The farmland didn't dissolve, it was added to ever-expanding corporate farm operations. Many rural towns had turned into ghost towns as the businesses that supplied farms and farm families closed. The US cheap food policy has not benefited the backbone of the countryside, the hardworking farmers and rural people who benefit from doing business with farmers.

The US cheap food policy that has driven US farms to increasingly higher levels of production to break even has come at a huge domestic cost. Although farm productivity has skyrocketed over the past five decades, farmers have not been justly rewarded. Traders like companies that sell farm chemicals and make processed food have taken the lion's share of profit. Farmers only get paid 14 cents of every consumer dollar spent on food, according to the USDA. But these 14 cents do not include the money farmers spend on inputs like seed and fertilizer, so farmers only pocket seven cents from every consumer food dollar.[231]

Former Secretary of Agriculture Tom Vilsack told the audience at the Farm Bureau Federation's annual convention in January 2022 that 437,000 farms have gone out of business since 1981, according to the latest census. In the same speech, he announced $200 million in loans and grants for the fertilizer and energy industries, a nod to the fact that special interests are clearly at the helm at the Department of Agriculture.[232] Agribusiness lobbyists position their corporations that trade in farm commodities as winners in the game of farm policy. Meanwhile, family farms have lost the autonomy to farm how they think is best for the land and their families and the American people have lost untold natural capital wealth in the form of soil degradation, shuttered communities, and shattered landscape function.

---

**229** United States Census Bureau, 1950 *Census of Agriculture: Volume 2. General Report* (1952), 5, https://www.census.gov/ library/publications/1952/dec/agriculture-vol-02.html.

**230** National Agriculture Statistics Service, *2022 Census of Agriculture Highlights: Farms and Farmland,* March 2024, 7.

**231** Michelle Miller, "A Breakdown of the Food Dollar: Where Does the Money Go?," *AGDAILY*, July 23, 2024.

**232** American Farm Bureau, "American Farm Bureau Convention Monday Recap: Remarks from Ag Secretary Vilsack, Ag Innovation Challenge and YF&R Awards, Farm Bill Congressional Discussion and More," news release, January 22, 2024,

One result of the Restoracene must be a re-peopling of farmland around the world so people can look after it. The best fertilizer is a farmer's footprint. An important part of holistic land management includes monitoring of ecological results. With larger farms and bigger GPS-guided equipment, human observations and insights are not guiding management. Many owners of US farmland live far away in cities and never get their fingernails dirtied by their soil. They only monitor profits in the reductionist approach that is washing America's once-vast wealth in fertile topsoil out to sea.

## Losing the Integrity of Our Land and Planet

As subsidized commodity crops prevail over the landscape, the decline of water infiltration rates means spring rains do not soak into soil and get stored as they should. Every springtime the Missouri and Mississippi Rivers flood communities who are so used to it they think it is normal. Bigger floods and droughts affect more citizens every year. Climate volatility will only make this worse in a tragic feedback scenario where great floods wash more soil away and make it ever harder for nature to protect us from the outcomes of our poor policy decisions. The way humans value yellow rocks exponentially more than water, the elixir of life, is a bizarre foible of humankind. But, if we stay on the same trajectory, water will become the new gold. It behooves us to learn how to enhance the water-absorbing, distributing, and storing ability of healthy ecosystems because the destruction of essential natural capital such as these undermines the Earth's life-support systems.[233]

Views of soil coursing through brown rivers after rain have greeted me in every state and country I have ever been. Pristine rivers are supposed to be clear unless they contain tannins. Then, they should look like weak tea and not like coffee with milk in it. People are so used to dirty waterways, even after just a little rain, that they don't even notice their national wealth in the form of topsoil being robbed from them until it inundates their house. Land growing monocultures of crops such as corn, wheat, and soybeans for food, fiber, and biofuels exhibit the highest rates of erosion.[234, 235]

233 William E Rees, "The Human Eco-Predicament : Overshoot and the Population Conundrum," *Vienna Yearbook of Population Research* 21, no. 39, (2021): 21–29. DOI:10.1553/p-eznb-ekgc21.

234 Pasquale Borrelli et al., "An Assessment of the Global Impact of 21st Century Land Use Change on Soil Erosion," *Nature Communications* 8, no. 1 (2017): 2013, https://doi.org/10.1038/s41467-017-02142-7.

235 David Pimentel and Michael Burgess, "Soil Erosion Threatens Food Production," Agriculture 3, no. 3 (2013): 448, https://doi. org/10.3390/agriculture3030443.

The US Corn Belt covers five states, from Ohio to Nebraska, and produces 75 percent of the nation's corn. In the Corn Belt, like grasslands around the world, centuries of plowing and overgrazing have resulted in catastrophic soil loss. Since settlers first put the plow to these once deep, dark prairie soils, the median soil erosion rate has been 14.3 metric tonnes per hectare per year.[236] Corn production causes the highest rates of soil erosion.[237] Compare this to the soil erosion rate from perennial grass of less than 1 tonne per hectare per year even in conventional grazing regimes.

## No-Till Is No Answer

No-till farming, where plants are sprayed with herbicides to kill them instead of plowing them under to prepare a seedbed, is not likely the solution chemical and fertilizer peddlers want you to believe it is. A meta-study looked at the long-term outcomes from no-till around the world over two decades. It found carbon in the whole soil profile down to 1.3 m showed absolutely no increase.[238] Rotating in several years of regenerative pasture is key to sustainable crop production because the fastest way to build soil is to mimic how nature did it: with animals living symbiotically with plants. Scientists have estimated that if all beef cattle in America were managed on pasture using adaptive managed pasture techniques, including during the finishing stage, the beef industry could build soil while sequestering 790 million tonnes of soil carbon per year.[239] This would be enough to offset all emissions from beef production in the US.

Countries that are producing far more crops than they need are forfeiting their national wealth to enrich global traders. Americans only use 8 percent of the corn they grow for food. The other 92 percent is used for animal feed, ethanol, and export, all of which enriches multinational agribusiness and fossil fuel industries. Annual monocrop farming today is hard on biodiversity, too, because it eliminates most other life forms via tillage, herbicides, and pesticides. On a global scale, reductionist decision-making around feeding our species has resulted in five billion hectares of man-made desert.[240] Row crop agriculture and overgrazing

**236** Evan A. Thaler et al., "Rates of Historical Anthropogenic Soil Erosion in the Midwestern United States," *Earth's Future* 10, no. 3 (2022): sec. 4, https://doi.org/10.1029/2021EF002396.

**237** David Pimentel et al., "Biofuel Impacts on World Food Supply: Use of Fossil Fuel, Land and Water Resources," *Energies* 1, no. 2 (2008): 52, https://doi.org/10.3390/en1010041.

**238** Neal R. Haddaway et al., "How Does Tillage Intensity Affect Soil Organic Carbon? A Systematic Review," *Environmental Evidence* 6, no. 1 (2017): 30, https://doi.org/10.1186/s13750-017-0108-9.

**239** W. R. Teague et al., "The Role of Ruminants in Reducing Agriculture's Carbon Footprint in North America," *Journal of Soil and Water Conservation* 71, no. 2 (2016): 160, Table 2, https://doi.org/10.2489/jswc.71.2.156.

**240** Walter Jehne and Tracy Frisch, "Supporting the Soil Carbon Sponge," *EcoFarming Daily*, June 20, 2019.

by pastoralists has released 116 billion tonnes of carbon from the soil through erosion and oxidation of carbon over the last 12,000 years.[241] Certainly, this cannot be sustainable. What kind of ancestors are we being to allow soil erosion to get worse year-on-year and put future food security at risk? I know from the decades required to grow soil on my farm, that replacing topsoil is not fast or easy. Are we putting the possibility of a livable planet at risk?

Every time soils are exposed to oxidation when they are plowed, more carbon is returned to the atmosphere. When we first bought Raindance Farm in the 1990s, we used to plow fields to plant new pastures. First, we would plant a mixed crop of sorghum and soybeans that the cows and wild deer could graze in the summer, followed by a cover crop of annual rye and hairy vetch in the autumn. In the following spring we would plant new pasture after discing our cover crop into the soil. Our soil organic matter never increased in those days, even with the cover crops.

Although we built soil over the 10 years of perennial pasture, our soil carbon percentage always went down after tillage. It wasn't until we stopped soil disturbance like plowing and harrowing, and adopted permanent, holistically managed pastures, did our soil organic matter begin to steadily rise. This makes sense because it was fully functioning ecosystems that built topsoil over eons. This made me realize that instead of using up and exporting the natural capital we require for life, people can restore fertile soil and aquifers while growing our food. Teaming up with the megafauna left on the planet, which are mostly made up of livestock now, will be our only option until wild populations can be restored.

The downward spirals of quality of life, natural beauty, equality, and our life-support systems continue to accelerate like a race car with the gas pedal stuck. But we have the choice to unstick it and avoid a ghastly future for our grandchildren. We can do much better, and we must. By mimicking nature, we can once again build natural capital wealth in the form of topsoil and clean fresh water. All it takes is being ecologically smart about how we capture solar energy.

## Hunger Has Its Roots in Inequality

The destruction of America's natural capital wealth in the form of topsoil loss continues today. Modern traders, including giant multinational agribusinesses, accumulate billions in wealth for their shareholders by exporting grains or turning them into unhealthy processed foods, while American farmworkers work long days

**241** Jonathan Sanderman et al., "Soil Carbon Debt of 12,000 Years of Human Land Use," *Proceedings of the National Academy of Sciences of the United States of America* 114, no. 36 (2017): 9576, https://doi.org/10.1073/pnas.1706103114.

in the field, often just scraping by. The adjusted net earnings of Archer Daniels Midland (ADM), which controls much US grain trading, processing, and benefits immensely from the subsidized ethanol industry, were $1.7 billion for 2024 alone.[242]

The most ubiquitous currency, money, can easily be used to launder socially unacceptable greed. You may pass a perfectly landscaped mansion and feel esteem for the residents you assume worked hard for the privilege of living there. If the owner had enriched himself in the illegal wildlife trade, you may never know and nobody would interfere. But, if the owner's great front lawn was piled high with dead, beheaded elephants and rhinos, it would be front-page news, eliciting outrage from the public. Neat and tidy monetary transactions can greenwash many sins, be they the exploitation of child labor, the dealing of drugs, the destruction of ecosystems in other countries, or the felling of rainforest.

Today, corporations have social and legal license to grow relentlessly while money greenwashes their greed. Globalization and unfair terms of trade allow people in wealthy countries, through commerce, to appropriate the natural capital of the people of poorer countries and the global commons such as the high seas. There is a resource grab occurring on our planet by wealthy countries who recognize that resources are dwindling. How can people know how much natural capital extraction is enough? Only by having an eye on the future welfare of our offspring and the health of our planet can decisions up to the highest corporate level be justly considered.

If farmers were provided with incentives to better steward the nation's natural capital, perhaps they could afford to plant more cover crops or put some cropland into pasture to restore its soils. But as profit margins are kept razor-thin and only certain row crops are subsidized by farm policy, conservation seems like an unaffordable luxury. The rise of global agribusiness and its cozy relationship with governments has allowed agribusinesses to profit from the depletion of the land and water resources of our citizens to enrich themselves in the name of the "feeding the world" myth. Colonial-mindset farming, the practice of seeking out pristine landscapes and erasing entire ecosystems to grow just one cash crop at the expense of local people, is alive and well today. Soybean production destroys precious rainforest in Brazil and palm plantations are replacing exquisitely diverse rainforest in Indonesia and Malaysia. It is time to reimagine a better agricultural system that is fit for purpose in the long term.

---

**242** Archer Daniels Midland, "ADM Reports Fourth Quarter and Full-Year 2024 Results; Provides 2025 Guidance," February 4, 2025, https://investors.adm.com/news/news-details/2025/ADM-Reports-Fourth-Quarter-and-Full-Year-2024-Results-Provides-2025-Guidance/.

## The Green Revolution

In January 2025, many Nobel and World Food Prize laureates collaborated on a letter calling for humanity to double down on technology to feed more people.[243] But they said nothing about the need to restore natural ecosystem productivity to help feed people, continuing to believe that human economies can exist apart from nature. Before we repeat history, we need to learn the lessons of the Green Revolution and how it worsened one of the root causes of hunger: inequality.

The Green Revolution began in the 1940s when hunger, caused by poverty and population growth in some nations outstripping their ability to produce enough food, was identified as a ticking time bomb for social unrest. Instead of addressing the root causes of hunger, people focused on technology. This had already been proven to damage the natural productivity of food-producing landscapes, such as the use of tillage, which had caused the Dust Bowl. Nevertheless, a push for new varieties of fertilizer-dependent monoculture crops and new ways to artificially fertilize and chemically treat these crops using fossil fuel-derived products created higher domestic food production in developing countries such as India and Mexico. But a study by researchers with the Public Health Foundation India highlights that increases in rice, wheat, and maize production came at the expense of a wide diversity of other indigenous crops and varieties. The chemical-intensive way crops are grown has damaged India's food-producing ecosystems so much that the researchers warn, "The green revolution, which was beneficial in ensuring food security, has unintended but harmful consequences on agriculture and human health. This requires new interventions to be tested and piloted before implementation, and continuous evaluation of the harms and benefits should guide the implementation."[244]

American farmers were told to accelerate the purchases of chemical inputs to "feed the world" but a result of the Green Revolution in America was an oversupply of crops that pushed domestic prices lower. Under the Agricultural Trade Development and Assistance Act of 1954, the US government decided to use food donations to dump the oversupply of corn and sugar they were producing on poor countries. In its 2004 Food Security report, the USDA admitted, "At that time, food aid was a feasible option for the disposal of rising US agricultural

**243** World Food Prize Foundation, "Hunger's Tipping Point: An Urgent Call to Transform Food and Nutrition Security," January 2025, https://www.worldfoodprize.org/index.cfm?nodeID=96854&audienceID=1

**244** Daisy A. John and Giridhara R. Babu, "Lessons From the Aftermaths of Green Revolution on Food System and Health," *Frontiers in Sustainable Food Systems* 5 (2021): Disc., para. 3, https://doi.org/10.3389/fsufs.2021.644559.

commodity surpluses."[245] But donating food aid to foreign countries had the unintended effect of lowering food prices in the recipient countries.[246] This disincentivized local farmers to grow their domestic supply and pushed many small local growers who fed their communities out of business, making food security more difficult to achieve.

The root causes of hunger, poverty and overpopulation, were not addressed by the Green Revolution. Instead of stamping out hunger, the additional food aid fueled unsustainable population growth, making the elimination of hunger ever more elusive. At the World Food Summit in 1996, Boutros Boutros-Ghali, then Secretary-General of the United Nations, pointed out the injustice of 800 million people still suffering from chronic malnutrition.[247] At that 1996 summit, the US government and other countries committed to reducing by half the number of hungry people worldwide by 2015. But the USDA Global Food Assessment in 2004 predicted that "the deep poverty at the root of hunger problems in the lower income population will remain unchanged."[248] In 2024, the USDA Food Security Assessment report revealed the total failure to cut hunger in half because still, "824.6 million people are unable to consume the 2,100 kilocalories (kcal) per day considered necessary for a healthy and active lifestyle."[249]

Green Revolution proponents often cite the increases in consumption (gross domestic product or GDP) experienced in poor countries as evidence of its success, but exploding populations in the most impoverished countries have negated any gains from increased GDP among ordinary citizens.[250] What would our world look like if root causes of hunger, such as poverty, overpopulation, and inequality, had been seriously addressed 60 years ago?

When women are better educated and less impoverished, birth rates decline.[251] What if everyone on the planet had the same access to education since the 1950s, all women had the economic and social wherewithal to make choices about family size, and the human population had remained stable at two and a half

---

**245** United States Department of Agriculture (USDA), Economic Research Service, "Food Security Assessment, GFA-15," (2004): 6, https://ers.usda.gov/sites/default/files/_laserfiche/outlooks/37133/30608_gfa15_002.pdf?v=86566https://ers.usda.gov/sites/default/files/_laserfiche/outlooks/37133/30608_gfa15_002.pdf?v=86566.

**246** USDA (2004), "Food Security Assessment," 7.

**247** Food and Agriculture Organization of the United Nations, *Report of the World Food Summit* (Food and Agriculture Organization of the United Nations, 1996), https://www.fao.org/4/w3548e/w3548e00.htm#Boutros

**248** USDA, "Food Security Assessment, GFA-15," (2004)" I.

**249** United States Department of Agriculture (USDA), "Food Security Assessment, GFA35," (2024): 2, https://ers.usda.gov/ sites/default/files/_laserfiche/outlooks/109853/GFA-35.pdf?v=72734.

**250** Rees, "The Human Eco-Predicament," 29–30.

**251** Jungho Kim, "Women's Education and Fertility: An Analysis of the Relationship Between Education and Birth Spacing in Indonesia," *Economic Development and Cultural Change* 58, no. 4 (2010): 739–74, https://doi.org/10.1086/649638.

billion instead of ballooning to eight billion? Would we still have almost a billion hungry people on the planet now that food production has quadrupled? If human population had been managed in a sustainable manner since 1960, could we be consuming 70 percent less fossil fuels and producing 70 percent less pollution and rubbish? Would much more rainforest still be standing and less fighting over resources be occurring across the globe? Could all people be content, well cared for, and happy in their home countries instead of so many people suffering?

William E. Rees, Professor Emeritus at the University of British Columbia, examined this question by measuring the ecological footprint of people since 1960, when the human population was 2.5 billion. He defines ecological footprint as the area of productive ecosystems required, on a continuous basis, to produce the renewable resources that the population consumes and to assimilate its carbon wastes. Between 1960 and 2017, the global human ecological footprint tripled from 7 billion to 20.9 billion hectares. The planet only has 12.1 billion hectares of productive ecosystems available, so we are in overshoot by almost another planet Earth. Population growth accounted for 80 percent of the increase in humanity's eco-footprint, far more than the increased consumption related to wealth.[252] This means the most powerful action people can take to write a better future for our grandchildren is having smaller families.

Sixty years after the start of the Green Revolution, a consensus of Nobel Prize-winning scientists warned about focusing on the symptoms of ecological overshoot such as climate change when human behavior was the problem, driven by economic growth, marketing, and pronatalism.[253] The insistence of other scientists to double down on technology instead of reimagining economies that do not reduce natural ecosystem productivity is an approach that will never improve the human condition.

## The Irony of Food Surplus

There is more irony in the story. People in rural areas are now the most food-insecure and have the fewest food choices. Our food choices have been reduced both in quality and diversity. More than 3.1 billion people—or 42 percent of the world's population—were unable to afford a healthy diet in 2021, but 1.5 billion people are overweight! This is far more than the 825 million people who went hungry in 2024.

---

**252** Rees, "The Human Eco-Predicament : Overshoot and the Population Conundrum," 28.

**253** Joseph J. Merz et al., "World Scientists' Warning: The Behavioural Crisis Driving Ecological Overshoot," *Science Progress* 106, no. 3 (2023), https://doi.org/10.1177/00368504231201372.

## The Efficiency Lie

Has the Green Revolution delivered food more efficiently? Not if you look at the food produced per unit of energy input. By that measure, agricultural energy efficiency has plummeted. In the 1930s, farmers produced more calories than they used. This is as it should be since farming can leverage the power of photosynthesis to capture solar energy. In 2009, modern food systems used 7.3 calories of fossil fuel energy to produce and deliver just one calorie of food.[254] As land degrades, more inputs, such as fertilizer and irrigation systems, are required to achieve the same level of production.

The proponents of the Green Revolution often point to the large surplus of grains the system produces, but the result has been widespread malnutrition because grains are, nutritionally, far inferior to animal products. A new meta-analysis found four billion people lack essential micronutrients, such as calcium, riboflavin, and vitamin B12, which are present in meat and milk.[255]

Even less logical is the growing and manufacturing of biofuels from corn, which uses 46 percent more energy in the form of diesel and chemicals (the production of which also requires energy) than it creates.[256] Biofuel production keeps farmers reliant on fossil fuels and agricultural inputs such as herbicides, pesticides, and fertilizers, ultimately enriching the global corporations that supply them. No doubt the production of biofuels is bleeding our nation's natural capital as soil washes away from cornfields, like I witnessed when I grew corn in Connecticut.

The most damaging unintended consequence of colonial mindset farming is the reduction of the fertility of the land we will leave for our grandchildren. While the Green Revolution temporarily helped reduce global hunger rates, the number of hungry people is on the rise again in some countries due to growing inequality and conflict. For instance, in 2023, Sudan exported $709 million worth of oily seeds, $693 million worth of sheep and goats, and $437 million worth of groundnuts (peanuts),[257] while 20.3 million Sudanese people went hungry, making Sudan one of the most food-insecure countries on the planet.[258]

---

**254** Michael Bomford and Richard Heinberg, *The Food and Farming Transition: Toward a Post-Carbon Food System* (Post Carbon Institute, 2009), 6.

**255** Simone Passarelli et al., "Global Estimation of Dietary Micronutrient Inadequacies: A Modelling Analysis," *Lancet Global Health* 12, no. 10, (2024): e1590, https://doi.org/10.1016/S2214-109X(24)00276-6

**256** Pimentel et al., "Biofuel Impacts," 51.

**257** OEC (The Observatory of Economic Complexity), *OEC Country Profile: Sudan* (OEC, 2025), https://oec.world/en/profile/ country/sdn.

**258** IPC, *Sudan Acute Food Insecurity Analysis: June 2023–February 2024* (IPC, 2024), 1, https://www.ipcinfo.org/fileadmin/user_upload/ipcinfo/docs/IPC_Sudan_Acute_Food_Insecurity_Jun2023_Feb2024_snapshot.pdf.

Due to worsening civil war and exploitation of resources by traders, from October 2024 to May 2025, more than half of Sudan's 46 million people (24.6 million) faced high levels of food insecurity.[259] But still, exports of over half a billion US dollars' worth of livestock and $293 million worth of groundnuts flowed out of the war-ravaged and hungry country in 2024 as part of its $3.13 billion in exports, mostly to the UAE, Saudi Arabia, and China.[260] These resources, along with oil and gold, are being plundered from the Sudanese people with the complicity of militias armed by foreign countries. The citizens had so little to show for the sale of their homeland's natural resources in 2024 that the usual imports of food like sugar and flour plummeted, contributing to the shrinkage of Sudan's normally high trade imbalance.[261]

## "Feeding the World" Makes It Harder for People to Feed Themselves

These exports come at a great cost to the present and future people of semiarid Sudan, as the largest aquifer on the planet is being drained to produce these crops for export, effectively exporting precious water that trickled underground when the Sahara was once a wet savanna. Again, history is repeating itself. Just like colonial English traders profited by exporting Irish-grown beef, butter, and barley to the UK during the Irish Potato Famine, traders today profit from the exports of millions of kilos of nutrient-dense protein while millions starve.

Great savannas that once covered much of the Arabian Peninsula and trickled water into aquifers have also been turned into desert at the hands of humans, so there are fewer functioning ecosystems for food production for Arab populations. Likewise, US fresh water is being raided for export at an accelerated pace, effectively robbing future American generations. Farmers in western US states squander rapidly depleting aquifer water to irrigate alfalfa fodder destined for mega dairy farms in the Middle East, some with up to 40,000 cows each. What will future generations think of how ours hurried to export on the cheap precious resources that include not only oil and natural gas, but the elixir of life itself, water?

**259** IPC, *Sudan: Acute Food Insecurity Situation: Updated Projections and FRC Conclusions for April 2024 to February 2025* (IPC, 2024), 1, https://www.ipcinfo.org/fileadmin/user_upload/ipcinfo/docs/IPC_Sudan_Acute_Food_Insecurity_Oct2024_May2025_Snapshot.pdf.

**260** Central Bank of Sudan, "Foreign Trade Statistical Digest: January 2024–December 2024" issue1, vol. 60, CBOS Policies and Research Assistance Department,9,18.

**261** Central Bank of Sudan, "Foreign Trade Statistical Digest: January 2024–December 2024," 6.

## Extinction Redemption

We can be better ancestors. It is beyond the imagination of most people to envision the return of fully functioning ecosystems that can replenish depleted aquifers around the globe, but this is exactly what thousands of holistic management practitioners like me believe can happen. After all, it was the biosphere that built our water and soil natural resources. Regenerative farmers with a different future trajectory in mind intend to reverse the negative effects of soil degradation and put vitality back into their soils. Once our waterways flow with crystal-clear water for weeks after rain, everyone will know regenerative farmers have achieved the outcomes they desire.

The good news is we don't need technology or highly trained scientists to solve the dual ecological crises of biodiversity loss and climate disruption. We only need to reconnect with nature and let her teach us more about the amazing green ecosystems that self-organize in the grace of her presence. We don't need a Western education or even to know how to read and write. We all can find grandmothers, grandfathers (either our own or someone else's), or local farmers and gardeners who know how to make our gardens, neighborhoods, and hometowns green. We can reconnect with people to leverage the resilient communities that can make your country, wherever you live, naturally productive and rich in natural capital again.

## Chapter 10

# The Raindance Farm Story

What is the best way to grow food? To heat our homes? To travel? With so much conflicting information in the media about what we should eat and what actions we should take to care for the planet, it is easy to feel overwhelmed. Should we eat meat or go vegan? There is no right answer because it depends on who you are, your culture, your needs, your budget, and what you want your future to look like. What is best will keep changing as the landscapes, climate, and circumstances change going forward. When you study holistic management through credible sources like the Savory Institute or Holistic Management International, you will learn to ask yourself questions that will help you decide which actions are right for you to bring you closer to the future you really want. Ongoing monitoring and replanning, if needed, makes sure improvements happen as conditions shift. After all, change is constant.

## Holistic Management

Holistic management (HM) was first developed over four decades by the Zimbabwean Allan Savory and other land management experts to reverse soil loss and desertification. Like most people, in the beginning I assumed HM was a grazing method and only for farmers. But, after practicing HM for a decade, I learned it was really about changing how we manage our lives, economies, and the natural world. Everyone who wants a better future can use this framework to make better decisions. It gets its name from the core principle that nature works in interconnected "wholes" where people, economies, and the environment are deeply intertwined.

In their book, *Holistic Management: A Common Sense Revolution to Restore Our Environment*, Savory and his co-author and wife, Jody Butterfield, begin by outlining the basics of ecological science and share some groundbreaking insights, such as how grazing animals play a role in preventing deserts from forming in semiarid environments. Then they offer a straightforward, step-by-step decision-making process to help people choose actions that are most effective in restoring lost natural capital, like topsoil and freshwater, while people also make a living. The holistic management framework is for city people as well as farmers, because everyone on the planet depends on healthy land where our food, oxygen, and water come from. Families can use holistic management to guide their buying decisions to increase the chances that their children and future generations will inherit a greener, kinder world.

HM urges us to focus on where we want to end up, not where we are. What I love best is Savory's concept of "holistic context"—the big game-changer. It gives individuals, policy makers, or businesses the chance to define how they want their lives to unfold in a sustainable society. You write your own future script first and then decide what your landscape must look like to provide what you need to live in such a way in that future. From then on, you make your decisions in the light of this guiding star, helped along by the step-by-step process embedded in the HM framework. In this chapter, I share the unique story of our holistic journey as an example of our why, our how, and the hope that our results inspired.

## Our Guiding Star at Raindance Farm

When my family and I created a holistic context for our farm, we first needed to define our "whole" under management: The decision makers included my family and our two employees. Our resources included 133 hectares of owned farmland, 20 rented hectares, the family members of our employees, and our customers. Our source of money was only the cash flow our farm operation generated.

Our guiding star holistic context at Raindance Farm was spelled out in two short paragraphs. The first succinctly described the quality of life we wanted for ourselves and our descendants seven generations from now:

> We live comfortably in energy-smart shelters where we enjoy peace and security. Like our neighbors, we have meaningful work and are fairly compensated. Abundant healthy food and clean water are available to everyone in our community. We have appealing choices and autonomy regarding our own well-being. We enthusiastically share culture with our neighbors. We have access to vibrant nature and ample leisure time to enjoy it all. We enjoy strong, cooperative connections with nature and our human community with whom we share mutual love and respect.

The second paragraph of our holistic context painted a picture of how our "resource base" needed to look to deliver that quality of life. In our case, our resource base was the land and waters within our farm boundary as described in what we defined earlier as our "whole," along with our relationships with people who were highly involved or did business with the farm. We tried to address the four ecosystem processes in this part: energy dynamics, water cycle, mineral cycle, and community dynamics. We spell out a picture of the context we want as if we were living in it:

> Our land supports a healthy, highly productive ecosystem enduring as a strong foundation for our food, energy, and fiber producing endeavors. It provides resources for a sustainable and reliable local economy and produces abundant healthy food for people and wildlife. Diverse and abundant flora and fauna function naturally in their historical ecological niches and evolution of well-adapted life forms is possible. Fertile and absorbent soils full of life are growing deeper. Fresh, clean water pours forth reliably from springs.

With our holistic context as our guide, we knew where to aim as we made all our important farm and family decisions as a team. We also could avoid a lot of debate and indecision each time. Since everyone had agreed to the holistic context, any action to move in that direction meant everyone was pre-boarded. For a family living in a town or city, the future resource base they need to look after might include their neighbors, the local watershed, the foodsheds where their food is grown, and the landscape in their neighborhood. The quality-of-life part of an urban family's context could reflect cultural or religious values. What matters is that everyone who is making management or consumer decisions shapes and owns the context.

The most potent lesson I learned after studying biology and anthropology to research this book was how similar we humans are. Human needs are, for the most part, universal, including the desire for love, peace, and security. Context creation is a powerful tool to bring people from different perspectives together. If feuding groups sit down and create a holistic context together with skilled guidance, they will find common ground rooted in our common human nature. Who doesn't want a better future for their children? This means people can be trained to use holistic management at the governance level, where so many legislatures are hobbled by opposition and take (or don't take) actions in a way that can make conditions worse for most people to benefit a few cronies on a shortsighted basis. With a holistic decision-making framework in place, entire communities, water districts, or companies can move toward the Restoracene while keeping economies running and bills paid. When I reinvented myself in New Zealand as a farm consultant with a new life partner, we created a holistic context for our household, a whole that overlaps with our sheep farm and the wholes of the other farm partners. In it, Rick and I have the resources and freedom to help make the world a better place—one of our core shared values and the reason I wrote this book.

## Our First Holistic Decision: Making a Farmstead Cheese

By 2007, after 10 years of organic dairy farming in New York State, we saw milk prices stagnate while costs climbed. Corporate entities had opened gigantic "organic" dairy factory farms out West. The Cornucopia Institute revealed with aerial photos that some of these concentrated feeding operations dishonestly failed to comply with even the minimum grazing rules set by USDA's National Organic Program (NOP). Aerial photography showed that 4,400 dairy cows at Aurora Organic Dairy's Colorado factory farm did not spend enough time outdoors to graze the minimum amount of pasture required by organic rules, so the Cornucopia Institute urged the USDA to investigate. The NOP found Aurora guilty of willfully violating 14 federal organic regulations and announced plans to revoke its organic certification.[262] But under the Bush Administration, Aurora only got a slap on the wrist. Despite mislabeling vast amounts of milk, the company stayed in business. The $100 million company was not even fined. In response, consumer groups from 40 states filed a class-action lawsuit and the Federation of Organic Farmers protested the USDA's leniency.[263]

Aurora was found to have used nonorganic cows to supply milk to major retailers like Costco, Wal-Mart, Target, and Safeway that was labelled organic. Honest organic dairy farms like ours, who raised their own cows from birth and fed them expensive organic milk and the best hay, were placed at an unfair disadvantage. Promiseland Livestock, a company that helped Aurora launder 13,000 conventional dairy replacements (young calves and heifers destined to become milking cows) as organic between 2004 and 2006, was not decertified until 2011.[264]

We had decided to go organic a decade earlier, so we could continue as a manageable-sized family farm instead of borrowing a lot of money to operate a large factory farm like many New York State dairy farms had needed to do because

**262** Mark A. Bradley to Marc Peperzak, April 16, 2007.

**263** Steve Morrison et al. to Chuck Conner, September 25, 2007, https://nodpa.com/files/2007-09-25_Letter_to_Acting_Secretary_Conner.pdf.

**264** Robert Vosburgh, "The Fall of Promiseland Livestock," Supermarket News, July 23, 2011.

prices were not covering costs. But the price premium designed to keep small organic family farms like ours afloat was eroding because mislabeled milk from factory farms was allowed to flood the market. It was getting increasingly hard to pay my bills even though my costs (except labor) were low compared to most dairy farms. I was happy to employ people and have them working on the land even if it meant my labor costs were higher than the average for my farm size in New York. I think the core purpose of any business should be to provide meaningful jobs. But I couldn't cash-flow my two employees with the prices we were offered.

## Checking against Our Guiding Star

We needed to adapt to remain in business. That's when we considered the idea of building a micro-creamery at Raindance Farm to add value to our organic milk by making cheese. After getting estimates for the cost of building a cheese plant on the farm, we ran the action of making a farmstead cheese through the set of holistic testing questions described in Savory's book to see if it would pass the context checks and move us toward the future we wanted:

- **Cause and effect.** Will building our own cheese plant address the root cause of the problem?
- **Social, biological, and financial weak links.** Socially, could farmstead cheesemaking create a weak link between us and the people we need on board? Biologically, would it address the weakest link in the life cycle of this organism? (This one did not apply in this case but could if people are considering spraying for pests, for instance). Financially, does this action strengthen the weakest link in the chain of production?
- **Marginal reaction.** When comparing all our options, does farmstead cheesemaking provide the greatest return toward the goal for each additional unit of time or money invested?
- **Gross profit analysis.** Which enterprises contribute the most to covering the overheads of the business?
- **Energy/money source and use.** Is the energy or money we will use to build and run the plant derived from the most appropriate source in terms of our holistic context?
- **Sustainability.** Will farmstead cheesemaking lead us toward or away from the future resource base described in our holistic context?
- **Gut feel.** After this analysis, do we still think farmstead cheesemaking will lead to the quality of life we want? Will it adversely affect the lives of others?

Since I felt the root cause of the problem was a milk price that was too low, this was easy to check off, although there were other ways to add value to our milk we had to consider. Then, we needed to discuss whether making farmstead cheese would cause a social weak link. There was a real risk of overworking everyone on the farm to undertake this project, but we realized that more stable prices could mean the difference between having jobs and selling the farm. Once the cheesemaking was up and running profitably in a couple of years, my financial spreadsheets showed how we could downsize our herd from 85 cows to 40 and still be profitable. This restructuring should make work easier and more enjoyable for my employees in the long run. From this point of view, they were on board with the idea.

Selling meat and cheese directly to our customers should strengthen, not weaken, our social bond with the people in our all-important market. We knew the taste of our milk was amazing when the cows were eating pasture. At farmers' markets, kids often pulled their parents over to our table to buy our cheese after they tasted a sample of the prototype cheese we had custom-made at The Cooperstown Cheese Company. The positive feedback we got from these happy customers fortified the meaningful connections with our community that we described in our holistic context.

A big part of creating a meaningful connection with our community was producing food that tasted great but was healthy, too. Because we only made Sun Cheese when the cows were on pasture, the conjugated linolenic acid (CLA) concentration it contained, 18 mg per gram of fat, was more than twice the concentration in conventional cheese on grocery store shelves. I suppose this was because we were meeting or exceeding the USDA requirements for pasture in our cows' diet. High pasture consumption is positively correlated with CLA levels. CLA has been linked to lower cancer rates and reduced obesity, which is ironic because it is a fatty acid.[265]

A fatty acid that prevented obesity made intuitive sense to me. When I visited traditional cheese-making regions in Europe as a girl, I saw only slender farmers walking hills to gather the cows in for milking. New research is confirming the health benefits of regeneratively grown food. A 2022 study showed food grown on regeneratively managed farms is more nutrient-dense than conventionally grown food, with higher levels of minerals and phytonutrients and healthier fatty

**265** Laura J. den Hartigh, "Conjugated Linoleic Acid Effects on Cancer, Obesity, and Atherosclerosis: A Review of Pre-Clinical and Human Trials with Current Perspectives," *Nutrients* 11, no. 2 (2019): 370, https://doi.org/10.3390/nu11020370.

acid profiles.[266] My son Dale loved to guzzle our raw milk right out of the jug in the fridge, which was probably a good thing since he didn't like vegetables as a teenager. He fortified his diet with fat-soluble vitamins and phytonutrients this way and grew big, strong, and lean. We were happy to produce healthy food for families and not just a commodity. Thus, farmstead cheese production passed the social weak link test because it would strengthen, not weaken, our all-important bond with our customers.

The financial weak link on our farm appeared to be a milk price that had stagnated while prices for the tools and materials we used had risen. There was little point in milking more cows to balance the books if the margins were poor. When we sold our milk on the fluid market, we had no control over the price. We were price takers.

Savory's third check, marginal reaction, asks decision-makers to compare the proposed action with doing nothing or something else to see which action provides the greatest return toward the goal for each unit of time and money invested. Farmstead cheesemaking in this case was a shaky proposition because the cost for the plant was very high per unit of cheese. We never planned on expanding production beyond the milk our farm could sustainably produce. But when we compared it to other ways to add value, like selling milk directly to customers or making butter, these would require even more money because we would need to buy a commercial pasteurizer. I considered the risk involved if we did nothing and we had to accept prices in the future that may not cover costs. The other thing to consider was the time required to design, build, and run a micro-cheese plant. It was hard to justify the cost since time is money, so this check was a fail.

When we looked at gross profit analysis, the milk enterprise was more profitable than the meat we sold. This made it the most logical place to invest in for adding value. On dairy farms, meat is a by-product, and we had several excellent butchers in our region who could add value to our meat without our needing to invest in any meat processing schemes.

The fifth context check, energy/money source and use, was interesting for me to explore. It asks, was the energy or money put into the cheesemaking project going to be in line with our holistic context? Because we were going to cash-flow the building of the plant ourselves by not downsizing our herd until we had the plant built, the source of our investment money to build the creamery came from

---

**266** David R. Montgomery et al., "Soil Health and Nutrient Density: Preliminary Comparison of Regenerative and Conventional Farming," *PeerJ* 10, e12848 (2022): table 4, https://doi.org/10.7717/peerj.12848.

the sun. Solar dollars work like this: Our land captured solar energy and grew plants that our cows turned into milk that we could sell to generate the funds.

We used a lot of creativity to make sure the energy used was also in line with our holistic context of leaving the world a greener place. We designed our cheese plant to be energy efficient. We repurposed the threshing floor of our quintessential 1880s Dutch barn for the cheese room to reduce the need to buy energy-intensive steel and other building materials. We incorporated energy-saving features such as plumbing the plant so we could pump our fresh warm milk directly from the milking parlor up to the cheese vat on the second floor, reducing the need to heat the milk for cheesemaking. We used solar-heated water for cleaning and to preheat the water that heated the cheese vat. We also installed a 13-kilowatt solar PV system with help from state and federal subsidies to provide our electricity.

Money and energy use seemed closely aligned with our holistic context and this also ticked the box of context check number six: sustainability. Farmstead cheesemaking would ultimately mean less exposure to milk price fluctuations, although it came with its own financial risks. Ultimately, it passed six of the seven holistic context checks including the last, gut feeling, and we decided to switch to on-farm processing.

## A Glimpse of the Restoracene

In our efforts to make healthy, great-tasting cheese, we brought so much wildlife back to our land. Despite the presence of predators like foxes and bobcats, we witnessed a massive increase in the populations of ground-nesting birds like endangered bobolinks—a native I had not noticed at all under the rotational grazing regime we used during our first 10 years on the farm. I think the shorter grass with our old management practices provided poor shelter for nests and we had been harvesting a greater area with mowing machines that destroyed the nests. A 2017 study confirmed that holistically managed pastures in New York State had up to 4.5 times more grassland birds than continuously grazed ones.[267]

In 2014, I was leading the cows in for their afternoon milking, and I noticed five rare Karner blue butterflies on a cow pie, of all places. This was the emotional moment when my dual lives as a lover of nature and a farmer became one. I felt their arrival was a personal thank-you note from the Creator, sent to encourage me to continue learning how to restore my land! Thirty years earlier,

---

**267** Lisa R. Cassidy and Gary Kleppel, "The Effect of Grazing Regime on Grassland Bird Abundance in New York State," *North-eastern Naturalist* 24, no. sp8 (2017): 91, https://doi.org/10.1656/045.024.0sp807.

when I was a college student, I had joined a protest in Albany, New York, over a plan to bulldoze and pave 70 hectares of a Karner blue butterfly habitat to build the giant Crossgates Shopping Mall. We lost that battle. Most initiatives that advocate for nature do fail. I felt pure joy all these years later to realize our family had found a way to restore habitat and grow food for these beautiful butterflies and countless other creatures, all while producing 350,000 liters of nutrient-dense organic milk for consumers each year.

## You Are What Your Food Eats

My favorite part of implementing the holistic grazing plan was seeing our cows become healthier. They shed their winter coats quicker, revealing shiny coats, and developed "happy lines"—horizontal bars across their bellies that indicated healthy rumen function. As their health and body condition improved, the incidence of lameness in our herd fell from 16 percent to 2 percent.

The average US dairy cow now lives four years before it is sold for hamburger meat, but our cows lived two or three times longer on average. I sold many of my older but still productive cows to Amish farmers who liked their grass-fed genetics from New Zealand and their "long handles"—older teats, ideal for hand milking. Because our cows were more fertile, we needed to raise fewer replacements—only 20 heifers per year instead of 28. This saved us $2,000 per calf for the cost of raising them, or $16,000 per year. The incidence of retained placenta our cows experienced at calving dropped from 8 percent to 1 percent. The number of mastitis cases per year between 2008 and 2015 dropped from seven to one on our farm. This was accomplished without antibiotics since we were a certified organic farm.

We learned how to identify the root cause of parasite outbreaks, and we eliminated dewormer treatments for our calves. We used none for the last six years I ran the farm. The calves were grazed every day like their mothers on fresh grass, which we never let get shorter than 15 cm. This kept their foraging muzzles out of the zone on the lower leaves of plants where parasite larvae live. We also let them have much more milk than before, either by suckling on their moms longer, or by weaning them after a full six months being fed whole organic milk. Our veterinary and medicine costs declined from $70 to $50 per cow per year. Mike Powers, our vet, even called me one day to ask why I wasn't calling him much anymore, worried I'd found another vet.

The presence of native brook trout in the stream running through the farm pastures told us we had preserved water quality, and the rain that percolated through our perennially shaded soils emerged from springs cool enough for this temperature-sensitive fish.

We saw a richer palette of plants in our pastures, including tall forbs like milkweed, the favorite food of monarch butterfly caterpillars. More wild legumes meant we did not have to buy as much fertilizer because legumes can fix nitrogen from the air and make it available to plants naturally. Forage grasses such as bromegrass (*Bromus spp.*), bluegrass (*Poa spp.*), and timothy grass (*Phleum spp.*) enhanced the taste of our meat and cheese, according to our customers at farmers markets and health food stores. They often thanked us for making cheese and sausages that their children loved. This deepened our connection with our customers in the positive way we had earlier envisioned in our holistic context.

Our progress toward our holistic context is summarized in Table 10.1. Except for two goals already in place at the start, we moved closer to the outcomes we had identified we wanted. Our pastures stayed greener through dry seasons, indicating deeper soil carbon and healthy life, though we wouldn't measure the full extent until years later.

**Table 10.1 Evidence of movement toward or away from our holistic context.**

| Element desired in our holistic context | Monitoring results relating to shift towards or away from context element after initiating holistic management | Outcome indicates a SHIFT TOWARDS context | Outcome indicates a SHIFT AWAY from context | Outcome indicates NO SHIFT |
|---|---|---|---|---|
| Landscape able to support profitable food protection | Farm profit continued to support two families milking 80 to 100 cows | | | X |
| Presence of abundant clean water | Pasture remained green longer during dry spells indicating improving water cycle and streams flowed crystal clear in flood events | X | | |
| Abundant quality food | Customers were willing to pay fair prices for our cheese and meat based on taste and the herd produced milk in the same abundance as before the trial | X | | |
| Rich wildlife diversity | Observations of species that were not observed before the trial: bobolink (*Dolichonyx oryzivorus*), bluebird (*Sialia sialis*), Karner blue butterfly (*Lycaeides melissa samuelis*), Coopers Hawk (*Accipiter cooperii*), Fisher cat (*Pekania pennanti*), bobcat (*Lynx rufus*) | X | | |
| Meaningful niche in society enjoyed by all family members and employees | Producing organic food for our customers was satisfying and appreciated by our customers | X | | |
| Healthy soil | Lifting trend in soil organic matter | X | | |
| Healthy animals | Reduction in mastitis 90%, reduced lameness in the cattle 88%, replacement rate dropped 18%, the need for parasite treatment reduced to zero | X | | |
| Abundant clean water | Brook trout present where grazing took place | | | X |

# Chapter 11

# Were We Carbon Farming?

There is new awareness of the need to sequester carbon to cool the planet and mitigate climate disruption that will otherwise cause untold damage to ecosystems and the humans living in them. Unfortunately, some foolish, expensive, resource-intensive schemes have been dreamed up to draw down carbon pollution from the atmosphere. For instance, carbon capture uses vast quantities of industrial equipment and pipelines that require fossil fuel energy to build. Planting pine plantations that don't sequester carbon long-term or selling more resource-intensive products like electric cars to undo wasteful pollution from fossil fuel burning is collective madness. Nature cools the global climate for free and always has. After I sold my farm and moved to the southern hemisphere, I met scientists who were experts in carbon sequestration who put cooling the planet into perspective for me. This understanding around the carbon cycle is of utmost importance for everyone to learn so we can prevent billions of dollars of your tax money and valuable time from being wasted.

So, why did I sell Raindance Farm and end up in New Zealand? For several years, I experienced unexplained chest pain. My doctor put me on asthma medications that didn't help. The lining around my lungs became inflamed with no explanation. In 2015, after a long day of tractor work, my left lung spontaneously collapsed. My doctor advised me to quit farming for the sake of my health. This was hard for me because my farm had become part of my identity. When walking down Main Street in Cooperstown, people would often stop me and say, "You are Raindance Farm! My kids love your cheese. Thank you." I cherished that kind of connection to my community. Luckily, I was able to sell the farm to an experienced farming family skilled in regenerative farming. We closed the deal at the beginning of 2016.

I returned the following summer to help on the farm while the new family dealt with a terminal illness. But first, I booked a long-overdue vacation to New Zealand. I had learned to surf at 15 and loved it, but I'd rarely had the chance to do much of it.

As a single mother of two since 2000 with a farm to run, there never seemed to be much time for vacations. By February 2016, I was happily exploring New Zealand's North Island surf breaks in a rental car, with a 5'10" surfboard in the back. Little did I know that scientists in the southern hemisphere would provide more clues that would help me understand what Raindance Farm was trying to teach me all along—how Mother Nature turns free energy from the sun into abundance. The efficient ecosystem templates nature created can help all people on the planet see paths to a kinder, greener future, so I needed to share my story of discovery.

## Green Country

In New Zealand, farmers are exceptionally skilled at pastoral farming and have great infrastructure to support grass-based farming systems. If you look at the country from Google Earth, you will see the farming areas are mostly clad in a protective green cover of pasture, making their agriculture much more sustainable than the often-bare soil of intensively cropped parts of the US, Canada, and Europe. I had copied New Zealand rotational grazing systems during my first 15 years farming, planting only perennial ryegrass and white clover and grazing the grass shorter. But the holistic planned grazing we started implementing in 2007 is not a rotational grazing system; it's an adaptive process that requires constant monitoring and adjustment based on ecological observation. The "secret sauce" of holistic management is the holistic context, the guiding star for all decisions. Without it, farmers are vulnerable to getting wooed by the next salesperson who comes up the farm driveway.

Although soil carbon under pastoral farms in New Zealand is far higher than in US cropping systems, it has not increased much in the last 20 years except on hill country grazed by beef cattle and sheep. The flat land where dairy farming and cropping is dominant has lost carbon, especially where swampland had been drained and converted to farming.[268] It is no coincidence that during my first 15 years of rotationally grazing my cows on perennial ryegrass pastures, my soil organic matter did not increase either. I think this was partly because I disturbed the soil with tillage to grow annual forages on five to ten percent of my farm every year before sowing new pasture. Another reason was that ryegrass and white clover bi-cultures require pasture recovery periods (length of time with no grazing) that are shorter for economical production. This means the plants and

**268** Louis A. Schipper et al., "A Review of Soil Carbon Change in New Zealand's Grazed Grasslands," *New Zealand Journal of Agricultural Research* 60, no. 2 ( 2017): 110, https://doi.org/10.1080/00288233.2017.1284134.

roots grew shorter. Plants do not share liquid carbon in the form of sugars with soil life very deeply when the roots only go 15 cm deep. It wasn't until I let my pasture plants grow taller in summer at Raindance Farm, and they put down deeper roots, that I saw a rise in soil organic matter. Then I discovered that ryegrass and white clover stay vegetative and nutritious even when they grow taller if they are part of a diverse pasture sward.

In 2018, I met Dr. Christine Jones while she was on a speaking tour in New Zealand. She is a brilliant Australian soil microbiologist, now retired, who had been researching liquid carbon exudates. She taught me that when plants are growing rapidly, they produce more carbohydrates via photosynthesis. Surprisingly, not all the chemical compounds resulting from photosynthesis (photosynthates) are used by the plant to grow material like leaves. Substantial amounts of photosynthates, like sugars and starches, are shared via the plant's roots with soil microbes that the plants have symbiotic connections with.[269] Fungi in the soil transport these carbon-rich liquid exudates along networks of filaments, bringing it wider and deeper in the soil profile and sharing it with other organisms. In return, minerals, water, and other nutrients flow back to the plants on the same fungal highways in nature's bustling underground economy.

## Green Pastures are the Key to Carbon Sequestration

I learned a lot by monitoring our holistic grazing plan over the years and taking a lot of notes. This way the farm became my teacher. In this section, I will share some of the insights I have encountered specifically for the benefit of people who take care of livestock and want to make their land more productive. Productive grasslands have more scope to be profitable, provide wildlife habitat, and sequester soil carbon.

Our experience of lifting soil carbon and pasture productivity per unit of fertilizer is good news for farmers who are interested in regenerative farming but are worried about how it will affect their bottom line. My story will reassure them that tall brown grass is not required to build a deep carbon sponge in temperate climates. The key is to focus on whole-ecosystem productivity, a major theme of this book and a desired outcome of holistic planned grazing as taught by Allan Savory.

---

**269** D. L. Jones et al., "Carbon Flow in the Rhizosphere: Carbon Trading at the Soil–Root Interface," *Plant and Soil* 321, no. 1–2 (2009): 6, https://doi.org/10.1007/s11104-009-9925-0.

The deeper plant roots go, the deeper carbon is pumped into the soil to feed soil-dwelling animals, insects, and microbes. Carbon shared by plant roots deeper in the soil profile is less likely to be consumed by bacteria for food and respired quickly back into the air. Much carbon has remained in some soil for hundreds and even thousands of years. This could explain why soil organic matter at Raindance Farm increased when we moved from basic rotational grazing and started holistic planned grazing. We focused on keeping our pasture green and growing fast to maximize pasture production by adjusting the recovery period we gave the pastures depending on the weather and seasons. Less overgrazing meant plants grew deeper roots and more photosynthesis occurred, so more carbon flowed into the soil to feed our livestock, both above and below ground.

Our grazing plan was indispensable for helping us keep our grass growing as fast as possible per unit of fertilizer input. It achieves this simply by optimizing the timing of our cattle movements. Pasture recovery period is the foundation of the Savory holistic grazing plan. When considering the range of pasture recovery times that would be best to meet both our environmental and financial objectives, we thought about the nutritional needs of the animals first because they were putting a roof over our heads.

Most farmers and ranchers move their stock from one pasture to the next when they run out of grass in the paddock the animals are in. Logical as it is, it often results in farmers shifting stock quicker when grass is growing slower because they run out of grass sooner. Unfortunately, when grass is growing slower the plants need more recovery time, not less. Suburban dwellers will intuitively understand this because people need to mow their lawns more often in wet spring weather than in dry summer weather. Also, landscapers often raise the mowing height of lawnmowers in the summer to prevent grass from drying out. We did something similar by raising the height of the grass we left behind after a graze (the residual pasture) in summer.

In nature, large wild grazing herds don't return to areas they have grazed until their dung has been processed by soil life and decomposed. Like people, all grazing mammals I know about do not like to eat near their own dung (except for rabbits who eat their own poo). When grass grows slower, such as when a dry spell comes or as winter approaches, the dung also decomposes slower because of less moisture or lower temperatures. So, on our grazing plan we had a longer recovery period target for our pastures at these times. Holistic grazing planning uses a brilliant but simple method of ratios to tell you how long your livestock need to be in each pasture to return in the recovery time you desire.

At Raindance Farm, our neighboring farmers would provide increasingly shorter pasture recovery times and then run out of grass if weather got dry and grass growth slowed down in the summer. How did we keep our cows well-fed and still provide the pastures with enough time to recover? This is where forward planning is indispensable. We needed to plan to either have extra area available to graze for times of slower growth, destock, or feed extra stored feeds like hay to give the plants the recovery time they needed and still feed the cows well. These options needed to be planned months ahead when we decided which pastures to delay grazing on or where we would make hay or baleage. If we waited until grass growth slowed, it was too late to have options in place that would keep recovery times sufficient to keep grass growing as fast as possible.

When we had a surplus of grass in May and June, instead of slowing down the rotation because we had more grass (and by default making the recovery periods longer), we took strategically selected pastures out of the rotation for that month to get extra rest. We delayed grazing them until later when grass growth slowed as summer approached and we needed more area for grazing.

By reducing the grazing area for the month, we could keep our cows moving rapidly through most of the pastures, but they would still harvest enough grass to chew down the sward to the residual height we wanted. This was also the time of year we increased their density by shifting them four times a day or so to trample anything they didn't eat, thus ensuring that quality would be excellent for the next grazing.

The delayed pastures grew deeper roots, and they could access water better in dry times. We tried to get them grazed while leaves were still green, and not yellow or brown, to meet the nutritional needs of our dairy cows. This timing usually worked well because the fields where we made first-cut hay or baleage needed more time to grow back than the pastures we had grazed because they were cut shorter. The delayed pastures gave us a place to graze the cows so the hay paddocks could finish recovering.

It was important for us to graze the delayed pastures while they were still green because Dr. Jones had taught me that when perennial grass plants go to seed, the flow of liquid carbon exudates slows down significantly. The pasture in the photo on the back cover of this book had been delayed by skipping it on the previous rotation, so it had a 47-day recovery time instead of the 22–25–day recovery times of the other pastures in May and June of 2015. We grazed it on July 3 because pasture growth was slowing down and we needed more area to graze.

I see livestock farmers in every temperate country I visit fall into the same trap of harvesting too much area for hay or baleage in the spring, and then not having enough grazing area available to their animals to rest pastures longer a month or two later when grass growth slows in summer. They are either forced to overgraze their too-small area of pasture or feed out the hay they just made, both of which are not cost-effective. Often, they simply overgraze their pastures by bringing animals back too soon before the plants they want are well-recovered, not realizing how much this is costing them in lower pasture yields and slowed carbon sequestration.

Livestock, like wild grazers, evolved to be selective eaters. There are thousands and even millions of years of wisdom embedded in their preferences. This is why we did not force our cows to eat everything in our pastures by leaving them in the paddocks too long. We instead allowed selective eating but tried to manage them to non-selectively trample the plants they didn't like so the resources they were using had an even playing field regarding competition. That is why we moved our cows to fresh grass at least four times a day in the springtime. It was important to establish high quality pasture early in the grazing season.

When we started holistic planned grazing, our pastures contained mostly perennial ryegrass and white clover, plants that produce a lot when grazed more often and to a shorter height. So, we gave our pastures just a little more recovery time every year in the summer, about the amount of time needed to grow an extra half of a leaf on the tillers of the grass plants that we wanted. In three years' time, we did indeed see a shift toward more palatable grasses like timothy, various legumes like red clover, and many forb species that take longer to fully recover. We discovered that very long rest periods favored the tall grasses that would have thrived on prairies through which migrating herds passed only periodically. If we wanted more flowering forbs and legumes, we used shorter pasture rest intervals to mimic pristine temperate open woodlands that would have had many resident grazing animals.

Many farmers in high-rainfall areas who are new to holistic planned grazing often let their pastures over-recover to go to seed. Instead, we timed grazing events to keep most of our pasture plants vegetative and growing quickly to keep our livestock and soil biology well fed. Our excellent soil test results indicated that tall grass is not required all the time to sequester a lot of soil carbon in a temperate climate like ours (800–1,000 mm of annual precipitation). About 15 percent of our pasture plants would slip through the cracks and not get eaten or trampled, creating a mosaic of plants at different stages. This mosaic produced more than enough seed to replenish our seed bank in the soil, and the taller clumps of grass were used by spiders to hang their webs and by birds to hide their nests. We didn't need to let a whole pasture go to seed on purpose because this would not have occurred in natural ecosystems that were replete with abundant grazers.

Milk cows, dry (non-milking) cows, and calves all had different nutritional needs that we tried to match by adjusting our grazing timing. We also considered the physical needs of the pastures when planning our grazing events, being mindful not to graze pastures when they were too wet and soft and could be damaged and compacted. If I knew a hurricane or big rain event was coming, I sometimes would graze flood-prone pastures even if they were not on the plan for that week and I would re-plan going forward to save well-drained pastures to graze after the rain event.

We moved the cows before they grazed pastures too short so the birds would have somewhere to nest. There were so many variables to consider that we could never have kept it all in our heads when we were faced with the daily decision of where to put the cows. Our grazing plan nicely organized all this complexity and helped us figure out the right timing for our cows to stay on any one paddock (grazing period) to get the results we wanted. One unexpected result was a significant increase in soil organic matter.

## Evidence We Lifted Soil Organic Matter

Like me, the new farm owners, John and Jana Meulenberg, also learned how to grow productive, high-quality pastures from holistic management educator Ian Mitchell-Innes. They have many years of experience with mob grazing and are good at it. The company they supplied milk to, Maple Hill Creamery, took baseline soil samples from all their supplying farms in 2017 to support their regenerative monitoring. At the Meulenbergs' farm, 39 soil samples (to a depth of 15 cm) were taken, covering all 197 grazable acres (89 hectares), and combined into a composite sample that was sent to the soil labs at Cornell University. The soil organic matter averaged 8.5 percent. This is 57 percent higher than the average soil organic matter on other New York State dairy farm pastures (5.4 percent) with silt loam soil types like ours.[270] A follow-up composite test five years later done on paddock 3 still showed soil organic matter at 8.5 percent, suggesting that the carbon levels had stayed high due to the continued holistic management.

---

**270** Joseph Amsili et al., *Characterization of Soil Health in New York State: Technical Report* (New York Soil Health Initiative, Cornell University, 2020), 16.

This higher carbon content matched what we saw on the land: Our pastures didn't dry out during summer droughts as they had in earlier years. We believed we had built a deep carbon sponge beneath the soil surface because rain soaked in quickly. Increasing soil carbon was linked to our desired holistic outcomes—extended periods of green growth and clean, abundant water.

We also had a useful baseline of soil measurements. In 2006, before we started holistic management, our fertilizer supplier took 60 soil samples (15 cm deep) from 8 hectares (20 acres) and combined them into three composite samples. The same lab found soil organic matter averaged 5.85 percent. This is very similar to a 2022 sample from a hayfield that was never grazed but managed organically (the "control" in (**Table 11.1** and **Figure 11.1**). During our years of rotational grazing in New York (between 1997 and 2006), we saw no significant increase in soil organic matter.

Pastures on our farm improved over time as our grazing management favored natural succession towards deep-rooted species such as red clover, orchard grass, and timothy. Four years after we started using holistic planned grazing, I was plowing a field to reseed it and noticed that the plants already growing in it were probably better than the plants I had planned to plant! That moment changed everything. I sold my plows, disc harrows, seed drill, and pasture seeder, and stopped all tillage on the farm. It was no longer needed and now I wouldn't be slicing through the life-forms that were building my soil, like earthworms and fungi.

When I used to grow only perennial ryegrass and white clover under a rotational grazing system, the pastures tended to "wear out." Sod-forming weedy grasses would creep into the pasture, and the pasture became less productive. The need to reseed my pastures occurred because of how I was managing the grazing all along by not allowing enough recovery time for the plants. Stopping all mechanical soil disturbance also lowered our fuel use and costs. The changes I saw above ground were mirrored by changes to the soil underground. **Figure 11.1** summarizes 15 years of soil organic-matter measurements at Raindance Farm. The lower values of samples taken from before we started holistic planned grazing (HPG) are consistent with the control measurement taken on the non-grazed hayfield area and the New York statewide average collated in a report on soil health in New York from pastures on similar soils from 2014 to 2018.[271]

**271** Amsili. (2020): 16.

**Table 11.1 Measurements of soil organic matter (SOM) from different pasture paddocks before and after we started holistic management.**

| Date | Trial | SOM (%) | Paddock |
|---|---|---|---|
| 12/15/2006 | Before | 5.4 | 4A - heifer |
| 12/15/2006 | Before | 6.25 | 9/11 |
| 04/17/2007 | Before | 5.9 | 3 (3E) |
| 08/11/2009 | During | 7.1 | 4A - heifer |
| 07/16/2013 | During | 9.6 | 30 |
| 07/16/2013 | During | 9.4 | 31 |
| 10/26/2017 | During | 8.5 | Whole trial composite |
| 07/15/2022 | During | 8.5 | 3 (3e) |
| 07/15/2022 | Control | 6.0 | Control |

## Carbon Farming

A 2 or 3 percent increase in soil organic matter doesn't sound like a big difference, but soil is very heavy. Try carrying around a big bucket full of soil for a while and you will see what I mean. According to The US Natural Resources Conservation Service Soil Survey, one square meter of our Chenango B class of soil typically weighs 1.35 tonnes (1.2—1.5 g per $cm^3$).[272] This means the top 15 cm (7 in.) of our soil on 1 hectare (2.47 acres) weighs a whopping 2,000 tonnes! If we increased the percentage of soil organic matter by 2.6 percent, that means we grew an additional 52 tonnes of water-absorbing soil organic matter on each hectare. Since organic matter is composed of about 58 percent carbon, this translates to an extra 30 tonnes per hectare of organic carbon added to the soil.

Translated into $CO_2$ equivalents, it represents 110 tonnes of $CO_2$ equivalent sequestered on each hectare of our farm with our switch to holistic planned grazing in just six years. If that occurred over the entire grazing area of 89 hectares, as the whole-farm testing results in 2017 support, then our cows helped our pasture pull

272 Leslie Crandall, Soil Survey of Chenango County, New York USDA Soil Conservation Service (1985): 179, Table 16, Chenango B.

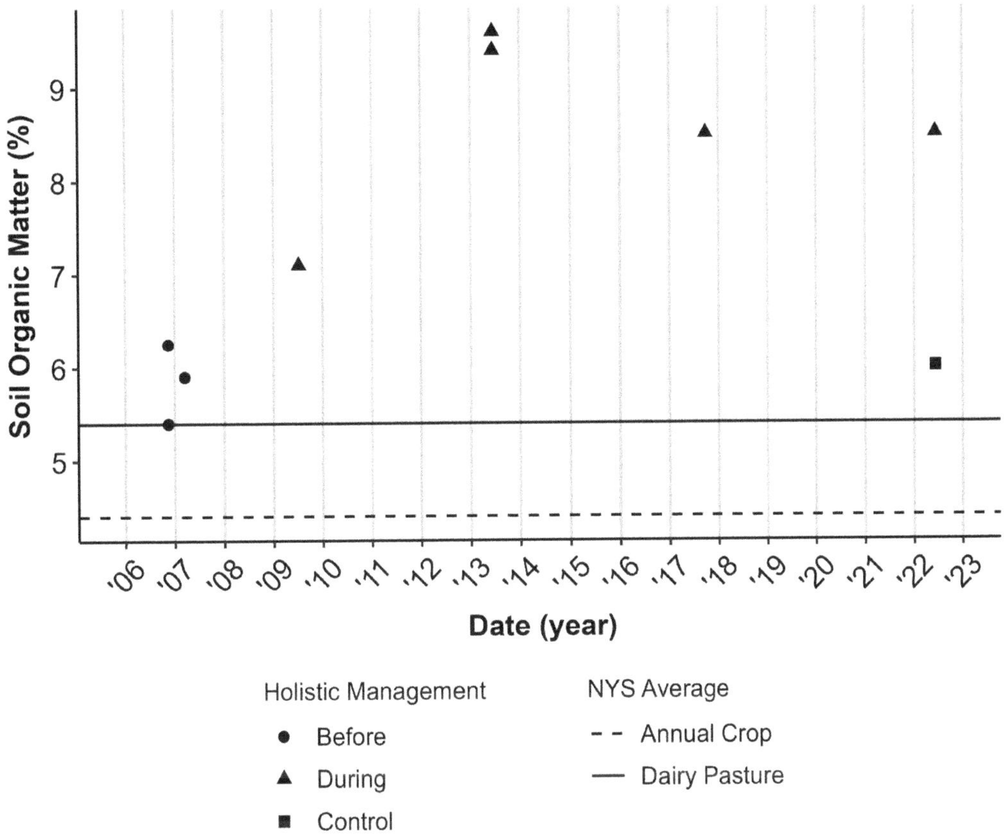


**Figure 11.1 Measurements of soil organic matter collected from cattle pastures on Raindance Farm over 15 years with average measurement from New York State soils of the same type.**

Created by Marcello de Andréa Segall.

at least 9,745 tonnes of $CO_2$ from the air and put the carbon it contained into the soil. This is likely a large underestimate because our soil aggregation and roots grew down to more than 100 cm in some fields. The soil test measurements of carbon used for the preceding calculations only measured the top 15 cm.

The carbon we sequestered between 2007, when we switched to holistic planned grazing, and 2017 translates to 10 tonnes of $CO_2$ equivalent per hectare per year! To put this into perspective, old-growth Amazon rainforest sequesters between 0.24 and 0.49 tonne of $CO_2$ equivalent per hectare per year in its above-ground biomass.[273] I don't mean to disparage rainforests. I love them.

**273** Jeffrey Q. Chambers et al., "The Steady-State Mosaic of Disturbance and Succession Across an Old-Growth Central Amazon Forest Landscape," *Proceedings of the National Academy of Sciences of the United States of America* 110, no. 10 (2013): 3952, https://doi.org/10.1073/pnas.1202894110.

The immense value of rainforests is as a home for tremendous biodiversity and their ability to produce climate-cooling rain. Likewise, the greatest advantage to society of increasing soil carbon at Raindance Farm is the inherent water storage ability of a healthy carbon sponge that reduces flooding and keeps plants green for longer, something that Australian scientists taught me was incredibly important for cooling the climate.

## Soil Carbon Matters More than We Think

Increasing soil carbon content is important because it helps restore the function of once-productive grasslands to stabilize climate through hydrology. My Australian mentor, Dr. Walter Jehne, taught me that water vapor is the most powerful greenhouse gas by far, responsible for 85 percent of the greenhouse effect, while $CO_2$ is responsible for only 11 percent. In the past, carbon-rich soil helped cool the planet by absorbing rainfall like a giant sponge. That moisture was then pulled up by plants and transpired into the air, feeding rainclouds and hydrating landscapes. In turn, green, moist landscapes covered and protected the soil with vegetation, reducing heating from solar radiation.

I remember walking barefoot on sun-heated sand to get back to my car after a morning surf session in Hawaii. I hopscotched from one green weed patch to another to keep my feet from burning. Now, imagine the solar heat radiated back into the atmosphere by the five billion hectares of bare ground across human-made deserts worldwide! In his presentation at the UK regenerative farming conference Groundswell, in 2024, Dr. Jehne explained that the heat reradiated by bare ground is subject to the Stefan–Boltzmann law.[274] This states that the total radiant heat energy emitted by an ideal heat-emitting surface is proportional to the fourth power of its absolute temperature (in Kelvin).[275]

Simply put, if the absolute temperature doubles, the heat radiated from an ideal absorbing and emitting material increases 16-fold. Thus, if dark soil exposed by deforestation in the Amazon reaches 70°C instead of the 20°C observed in the shade of trees or grass, the bare soil will reradiate far more heat into the atmosphere than you might expect. In the atmosphere, greenhouse gases can trap the heat. Cooling the planet isn't just about carbon; it's about preventing a lot of the sun's energy from reradiating from bare surfaces like concrete and bare soil.

---

**274** Walter Jehne, "Harnessing Nature's Solutions to Grow Soil," address to the attendees of Groundswell UK, July 1, 2024, 6:20–7:20. https://regenerate-earth.org/walter-jehne-at-groundswell-uk-2024/.

**275** Britannica Editors, "Stephan–Boltzmann Law," https://www.britannica.com/science/thermal-radiation.

To cool the landscape where you live, people can vegetate city and town highway verges, parking lots, yards, and playgrounds near them by planting trees, shrubs, and flowers to help cool their local landscapes. But the biggest opportunity to cool the entire planet is if humans regreen some of the five billion hectares of land that has turned to desert in the last 50,000 years.

## The Carbon Sequestration Potential of Landscape Function

Holistic planned grazing over 16 years on our farm helped restore biodiversity and soil carbon. I am not alone in using holistic management to restore natural systems. As farmers move toward an abundance of life and healthy soils, they can help cool the climate by increasing soil carbon.[276, 277, 278, 279, 280] If farm policy in the US encouraged widespread uptake of regenerative farming and ranching, food production could be carbon-negative with no need to reduce livestock numbers because the animals would drive carbon sequestration.[281] Consumers can have a say in this potential outcome every day. By seeking out locally sourced, regeneratively grown foods, families can vote for more climate-cooling life on their foodscapes.

**276** Adam Sacks et al., "Reestablishing the Evolutionary Grassland–Grazer Relationship to Restore Atmospheric Carbon Dioxide to Preindustrial Levels," in *Geotherapy* (CRC Press, 2014), 185–86, https://www.taylorfrancis.com/chapters/oa-edit/10.1201/b13788-9/ reestablishing-evolutionary-grassland%E2%80%93grazer-relationship-restore-atmospheric-carbon-dioxide-preindustrial-levels-adam-sacks-richard-teague-fred-provenza-seth-itzkan-jim-laurie?context=ubx&refId=1ce01fd6-2f18-4718-8d93-96e3d4c2a9f5.

**277** Richard T. Conant et al., "Land Use Effects on Soil Carbon Fractions in the Southeastern United States. I. Management-Intensive versus Extensive Grazing," Biology and Fertility of Soils 38, no. 6 (2003): 390–91, https://doi.org/10.1007/s00374-003-0652-z.

**278** Rattan Lal, "Conceptual Basis of Managing Soil Carbon: Inspired by Nature and Driven by Science," *Journal of Soil and Water Conservation* 74, no. 2 (2019): 32A, https://doi.org/10.2489/jswc.74.2.29A.

**279** Priscilla K. Lalampaa et al., "Effects of Holistic Grazing Management on Milk Production, Weight Gain, and Visitation to Grazing Areas by Livestock and Wildlife in Laikipia County, Kenya," *Ecological Processes* 5, no. 1 (2016): 17 https://doi. org/10.1186/s13717-016-0061-5.

**280** David Yarrow, "Geology into Biology: Carbon, Minerals, and Microbes—Tools to Remineralize Soil, Sequester Carbon, and Restore the Earth," in *Geology*, (CRC Press, 2014), 221, https://doi.org/10.1201/b13788-10.

**281** W. R. Teague et al., "The Role of Ruminants in Reducing Agriculture's Carbon Footprint in North America," *Journal of Soil and Water Conservation* 71, no. 2 (2016): 161-62, https://doi.org/10.2489/jswc.71.2.156.

# Chapter 12

# Ocean Armageddon

The raindrops that fell on our farm soaked into the soil and percolated through the ground over days or weeks before reemerging, cleansed, from springs. This water supplied streams that supported fish and amphibians like frogs and salamanders. These streams in turn joined forces with others to form the Cherry Valley Creek that merged into the mighty Susquehanna River, which meets the Atlantic Ocean at Chesapeake Bay. It was important that water left our farm clean because many towns downstream from us in the Susquehanna watershed drew their drinking water from the river. Ensuring that clean water leaves all planetary landscapes on its way to our oceans is the first step for ocean restoration. But why is ocean restoration so important for a livable future? What can you do to save the wildlife in our oceans that keep it healthy?

When I witnessed how holistic management helped us improve biodiversity, productivity, and soil carbon on our land, I was hopeful that restoring landscapes could be the key to a more stable climate. While it is a good start, as I delved deeper into carbon cycle research, I uncovered a surprising truth: deep-ocean sediments provide significant long-term storage of carbon.

Kickstarting massive carbon sequestration in deep-ocean sediments is very easy for people to do in principle: Give the oceans a rest from industrial fishing and pollution to allow the marine animals that drive carbon sequestration to thrive. But the reductionist management that is occurring around our oceans is resulting in a precipitous crash of wildlife populations that mirrors the carnage on land. Reductionist management can never tackle the complex task of protecting the integrity of our oceans and succeed. Only a holistic approach by everyone on the planet will be adequate to save highly evolved and intelligent ocean wildlife and help us write a better future for people.

Why should we care? Because ocean life absorbs half of all the atmospheric $CO_2$ drawn down by biology through photosynthesis.[282] This means it provides half of the oxygen we breathe. Ocean life must thrive for these ecosystem services to continue. Oceans contain 50 times more carbon than the atmosphere, an astounding 38,000 gigatonnes, and both are intrinsically connected.[283] Restoring life in our oceans can have a profound positive impact on biodiversity and climate. Rewilding fish has been described as the best climate solution among nine animal-led strategies.[284] Unlike the hands-on labor needed to restore ecosystems on land, restoring oceans involves less investment from us. We just need to give them a break. But like ecosystem restoration on land, this is not so simple to achieve when eight billion people need to eat.

## The Problem Beneath the Waves

Many aquatic species are fighting for their survival in ecosystems that have been rapidly degraded by people in the last couple of centuries. Due to "shifting baseline syndrome," each generation assumes the feebler oceans they inherit are normal, but they contain nothing close to the abundance nature had amassed as recently as 100 years ago.[285] When I snorkel near my home on New Zealand's South Island, I notice how few fish, seabirds, and mammals I see. It reminds me of Rachel Carson's *Silent Spring*, but underwater. A local elder, Ian McDonald, tells me stories about his visits to the same seaside when he was young in the 1950s and '60s. He and his friends were guaranteed to come home with a feed of blue cod just by fishing off the rocks with a hand line. They didn't even need a fishing rod.

The awful truth is that industrial fishing and associated bycatch, where non-target species caught in nets are dumped dead and dying back into the ocean, have stripped our oceans of most of their fauna in recent decades. The biomass of large, predatory fish like tuna has declined by two-thirds over the last century, with more than half of it occurring in just the last 40 years. Compared to pristine conditions, marine mammals are one-fifth their original biomass.[286]

---

**282** Michael J. Lutz et al., "Seasonal Rhythms of Net Primary Production and Particulate Organic Carbon Flux to Depth Describe the Efficiency of Biological Pump in the Global Ocean," *Journal of Geophysical Research* 112, no. C10011 (2007): 1, https://doi.org/10.1029/2006JC003706.

**283** R. A. Houghton, "Balancing the Global Carbon Budget," *Annual Review of Earth and Planetary Sciences* 35, no. 1 (2007): 315, https://doi.org/10.1146/annurev.earth.35.031306.140057.

**284** Oswald J. Schmitz et al., "Trophic Rewilding Can Expand Natural Climate Solutions," *Nature Climate Change* 13, no. 4 (2023): 328, https://doi.org/10.1038/s41558-023-01631-6.

**285** Masashi Soga and Kevin J. Gaston, "Shifting Baseline Syndrome: Causes, Consequences, and Implications," *Frontiers in Ecology and the Environment* 16, no. 4 (2018): 222, https://doi.org/10.1002/fee.1794.

**286** Yinon M. Bar-On et al., "The Biomass Distribution on Earth," *Proceedings of the National Academy of Sciences of the United States of America* 115, no. 25 (2018): 6508. https://doi.org/10.1073/pnas.1711842115.

By 2002, adult cod fish biomass in the North Atlantic was about 3,000 megatonnes, which is less than 0.3 percent of the biomass of adult cod in 1852.[287] This is a decline of three orders of magnitude that should alarm everyone. These outcomes are proof that fisheries are not being managed sustainably, despite marketing claims. It is a bigger problem than most people think because the interaction between aquatic living organisms can have disproportionately large indirect effects on the biogeochemical processes that allow carbon to be sequestered in deep ocean waters as dissolved inorganic carbon.[288]

This tragedy extends to keystone species like haddock, sharks, and whales. Blue whales, the planet's largest animal, now exist at under 1 percent of historic levels.[289] Seabirds like majestic albatross are also being affected by industrial fishing. They are being caught by accident on fishing long lines and are declining because of less food due to overfishing.[290]

Not only does this overexploitation reduce the natural capital of the commons in the form of fish for food, but the climate ramifications are severe. When fish biomass is reduced, ocean productivity falls because fish, like animals on land, cycle nutrients and fertilize plants and algae. Pristine ocean fish biomass had been reduced 50 percent due to fishing by the time fish harvests peaked between the 1960s and the early 2000s.[291] This reduction in productivity weakens a natural, fish-powered mechanism for cooling the planet.

## Tiny But Powerful Solar-Energy Converters

Phytoplankton, tiny one-celled organisms, power ocean productivity by capturing the energy of the sun and turning it into chemical energy that feeds the ocean. They form the foundation of marine food webs like plants do on land. Alarmingly, the primary productivity of the oceans as measured by phytoplankton abundance has declined at an average rate of 1 percent per year since 1899.[292]

---

**287** A. A. Rosenberg et al., "The History of Ocean Resources: Modeling Cod Biomass Using Historical Records," *Frontiers in Ecology and the Environment* 3, no. 2 (2005): 88, https://doi.org/10.1890/1540-9295(2005)003[0078:THOORM]2.0.CO;2.

**288** Oswald J. Schmitz et al., "Animating the Carbon Cycle," *Ecosystems* 17, no. 2 (2014): 345-6, https://doi.org/10.1007/ s10021-013-9715-7.

**289** Angela L. Sremba et al., "Circumpolar Diversity and Geographic Differentiation of MtDNA in the Critically Endangered Antarctic Blue Whale (Balaenoptera Musculus Intermedia)," *PLOS ONE* 7 (3): e32579 (2012): Intro., https://doi.org/10.1371/ journal.pone.0032579.

**290** Deborah Pardo et al., "Additive Effects of Climate and Fisheries Drive Ongoing Declines in Multiple Albatross Species," *Proceedings of the National Academy of Sciences of the United States of America* 114, no. 50 (2017): e10836, https://doi. org/10.1073/pnas.1618819114.

**291** Daniele Bianchi et al., "Estimating Global Biomass and Biogeochemical Cycling of Marine Fish with and Without Fishing," *Science Advances* 7, no. 41 (2021): 3, https://doi.org/10.1126/sciadv.abd7554.

**292** Daniel G. Boyce et al., "Global Phytoplankton Decline over the Past Century," *Nature* 466, no. 7306 (2010): 591, https://doi. org/10.1038/nature09268.

Ocean degradation follows the familiar pattern we observed on land. First, humans removed megafauna such as cod, whales, and sharks by hunting them to near extinction levels. Then people hunt their way down the size gradient to smaller species until the ecosystems lose their ability to feed us sustainably. Now even tiny krill, food for a plethora of species including fish and baleen whales, are being hoovered up from the oceans by fishing fleets.

The lessons I learned practicing holistic management at Raindance Farm—about biodiversity, nutrient cycling, and productivity—apply equally to the sea. We need to manage our oceans holistically, too, if we want to reverse ocean degradation so that we can benefit from the food abundance and climate cooling ecosystem services oceans provide. The good news is that it is much easier to restore ocean life; we just need to apply the management tool of rest.

## Productive Oceans Cool the Planet

Applying the tool of rest to restore preindustrial levels of marine life would help reduce atmospheric greenhouse gases in three ways: by lowering emissions from reduced fuel use by fishing fleets, by allowing biomass that contributes to long-term carbon storage to increase, and by allowing nutrient cycling to speed up to sequester carbon faster in living and dead biomass.[293]

Megafauna on land transport nutrients from areas of high fertility to less fertile areas, making ecosystems more productive.[294, 295] The same occurs in the sea. Whales and pelagic fish move nutrients laterally around the ocean, making it more productive.[296] This lateral movement of nutrients mirrors the way diprotodons, mammoths, and bison once boosted productivity on land beyond anything we can imagine today in ecosystems that have been impoverished by many generations of people. Just like on land, when ocean ecosystems are more productive, they use and store more carbon, keeping it out of the atmosphere with a cooling impact upon global climate.

---

**293** Gaël Mariani et al., "Let More Big Fish Sink: Fisheries Prevent Blue Carbon Sequestration—Half in Unprofitable Areas," *Science Advances* 6, no. 44 (2020): 3, https://doi.org/10.1126/sciadv.abb4848.

**294** Adam Wolf et al., "Lateral Diffusion of Nutrients by Mammalian Herbivores in Terrestrial Ecosystems," *PLOS ONE* 8, no. 8, e71352. (2013): 9, https://doi.org/10.1371/journal.pone.0071352.

**295** Christopher E. Doughty et al., "The Legacy of the Pleistocene Megafauna Extinctions on Nutrient Availability in Amazonia," *Nature Geoscience* 6, no. 9 (2013): 761, https://doi.org/10.1038/ngeo1895.

**296** Joe Roman et al., "Whales as Marine Ecosystem Engineers," *Frontiers in Ecology and the Environment* 12, no. 7 (2014): 380–81, https://doi.org/10.1890/130220.

## The Whale Pump

The average whale can sequester 13 tonnes of $CO_2$ equivalent as carcass "deadfall," which on its own helps sequester carbon. But even more massive ecosystem benefits could be harnessed by helping whales recover to their original populations because of their fertilizing effect on phytoplankton when they are alive and interacting with their environment. Whales also move nutrients vertically in the water column. In the Gulf of Maine, where nitrogen is a limiting factor, whale activity has been found to drive ecosystem productivity.[297] By feeding in deep waters and defecating near the water surface when they come up to breathe, they fertilize phytoplankton, the primary producers in this ecosystem.

Phytoplankton themselves sink down to deep ocean storage, but more importantly, they are the foundation of the ocean's food pyramids. So, more phytoplankton means more fish and marine mammals if they are protected. Warm tropical waters are low in nutrients but receive abundant light, so here the fertilizing effect of whales is more pronounced as they naturally translocate nutrients from their cold feeding grounds to tropical waters.[298] Fish also excrete highly available nitrogen, a little-known biological mechanism that supports rapid phytoplankton growth and carbon drawdown.[299]

## The Biological Carbon Pump Powered by Fish

About one-fifth of biomass ingested by fish forms fecal pellets, an important source of large organic particles that sink faster than smaller particles like phytoplankton clumps or krill pellets. A percentage of these ends up in deep ocean sediments. The carbon within these pellets and the carcasses of the fish themselves can remain stored in sediments far from the atmosphere for hundreds to thousands of years. Yale researchers estimate the net carbon sequestration effect of fish is at least 1.5 gigatonnes per year.[300] By comparison, the net carbon sequestered by the entire Amazon forest is 0.1 gigatonnes of $CO_2$ equivalent per year.[301]

**297** Joe Roman and James J. McCarthy, "The Whale Pump: Marine Mammals Enhance Primary Productivity in a Coastal Basin," *PLOS ONE* 5, no. 10, e13255 (2010): Results and Disc., https://doi.org/10.1371/journal.pone.0013255.

**298** Joe Roman et al., "Whales as Marine Ecosystem Engineers," 380–81.

**299** Callum M. Roberts et al., "Marine Reserves Can Mitigate and Promote Adaptation to Climate Change," *Proceedings of the National Academy of Sciences of the United States of America* 114, no. 24 (2017): 6170, https://doi.org/10.1073/ pnas.1701262114.

**300** Grace K. Saba et al., "Toward a Better Understanding of Fish-Based Contribution to Ocean Carbon Flux," *Limnology and Oceanography* 66, no. 5 (2021): 1644, table 2, https://doi.org/10.1002/lno.11709.

**301** Nancy L Harris et al., "Fluxes," Nature Climate Change 11, no. March (2021): 235, https://doi.org/10.1038/s41558-020-00976-6.

There is a strong positive correlation between ocean productivity and carbon sequestration from the ocean surface to deep-sea sediment.[302] But because the biomass of fish targeted by fishing has been reduced by half, scientists estimate 30 percent fewer fish pellets sink to the deep ocean for long-term sequestration.[303] Protecting marine life through regenerative fishery management or rewilding in marine reserves could accelerate atmospheric $CO_2$ removal by speeding up ocean nutrient cycling and productivity.[304] Ending industrial fishing subsidies and prioritizing marine restoration could protect this essential function. Fish biomass can potentially increase once intense fishing pressure is relieved, offsetting more significant human emissions if humans can collectively muster the fortitude to restore our oceans.

Ocean ecosystem models have shown that restoring the populations of large generalist predators like sharks or gigantic secondary consumers like whales to a "pristine" state could increase total ecosystem biomass by 267 percent.[305] This would more than double the powerful biological carbon pump. But can we multiply the power of the biological pump even more by setting the bar of abundance higher in the fisheries we manage? The successful marine reserve at Cabo Pulmo, Mexico, proved that fish populations can rebound quickly, given the chance. The biomass of fish in the Cabo Pulmo Marine Park increased 463% in the first decade after its establishment.[306] For the young people today who want a comfortable climate and abundant food as part of their future holistic context, advocating for marine protection will go a long way.

## Neptune's Secret to Achieve Abundance

Nature loves to evolve toward the most efficient use of solar energy. In the oceans she can get the wheels of productivity spinning incredibly fast, with huge ramifications for global climate. The ratio of plants to consumers in terrestrial food pyramids is 23:1 because there is so much plant matter available to feed animals that consume it.[307] Counterintuitively, in oceans, the distribution of biomass between plant and algal primary producers and consumers like fish is

---

**302** John P. Dunne et al., "Empirical and Mechanistic Models for the Particle Export Ratio," *Global Biogeochemical Cycles* 19, no. 4 (2005): sect. 6.3, https://doi.org/10.1029/2004GB002390.

**303** Bianchi et al., "Estimating Global Biomass," 5, *Disc.*

**304** Roberts et al., "Marine Reserves," 6170.

**305** C. Brock Woodson et al., "A Unifying Theory for Top-Heavy Ecosystem Structure in the Ocean," *Nature Communications* 9, no. 1 (2018): 23, https://doi.org/10.1038/s41467-017-02450-y.

**306** Octavio Aburto-Oropeza et al., "Large Recovery of Fish Biomass in a No-Take Marine Reserve," *PLOS ONE* 6, no. 8, e23601 (2011): Results, para. 2, https://doi.org/10.1371/journal.pone.0023601.

**307** Yinon M. Bar-On et al., "The Biomass Distribution on Earth," fig. 2.

inverted at 1:5. In other words, the trophic food pyramid is turned upside down as consumers outweigh producers by five to one. How is this possible?

The secret lies in higher productivity, the holy grail of all ecosystems, allowing phytoplankton to grow and reproduce at a turbocharged rate. Computer modeling suggests hourglass-shaped trophic pyramids, where the total biomass of the largest consumers and predators outweighs that of smaller consumers, was likely normal for pristine marine ecosystems and can be possible in some terrestrial ecosystems such as in Africa, where gigantic animals like elephants are present.[308] This occurs because the larger animals live longer and have more efficient metabolisms. This could explain why Alfredo Girón, the Head of Ocean Action Agenda at World Economic Forum, recounts how sharks and other apex predators have rebounded tenfold under protection in the Cabo Pulmo Marine Park.[309] This is encouraging news for people today who want to reimagine much more abundant landscapes and fisheries in the future. It means if people restore ocean life, numbers of large predatory fish like tuna and great mammals may increase far more than we ever imagined.

The additional ocean biomass accumulating in the bodies of the great diversity of whales, sharks, and other large marine animals increases the potential for long-term carbon sequestration. Scientists have estimated that restoring the populations of just eight species of whales would store 8.7 million tons of carbon in their bodies, equivalent to preserving a forest the size of Rocky Mountain National Park. As more tonnage of whale carcasses and fecal pellets descend to deep ocean carbon sinks, it would sequester an additional 160,000 tonnes of carbon per year to deep-sea sediments, where it could sit for thousands of years.[310]

On land, resting ecosystems from grazing doesn't always give positive outcomes, especially in semiarid climates. But in the ocean, the management tool of rest for overfished areas can achieve remarkable results, allowing fish populations to recover. Over time, this can improve fishing for nearby communities as well. British environmental journalist Charles Clover has documented many instances of this occurring in marine reserves around the globe and writes, "To rewild the sea is to bring back lost and depleted species to our oceans and restore ecosystems that have been harmed by human activities—simply by

**308** C. Brock Woodson et al., "A Unifying Theory for Top-Heavy Ecosystem Structure in the Ocean," *Nature Communications* 9, no. 1 (2018): 23, https://doi.org/10.1038/s41467-017-02450-y.

**309** World Economic Forum, Nature and Biodiversity, "A small Mexican town rallied to revive its overfished coral reef. Here's how," Jun 19, 2025, Video, min 1:27, https://www.weforum.org/stories/2025/06/community-ocean-conservation/

**310** Andrew J. Pershing et al., "The Impact of Whaling on the Ocean Carbon Cycle: Why Bigger Was Better," *PLOS ONE* 5, no. 8, e12444 (2010): 4, Disc., https://doi.org/10.1371/journal.pone.0012444.

stepping back and letting nature repair the damage, or by reintroducing species or restoring habitats."[311]

## Inequality on the High Seas

Industrial fishing fleets have disempowered local artisanal fishers, sweeping up vast amounts of fish on the high seas and stealing fish from the territorial waters of other countries. My daughter and I witnessed the effects of this firsthand in 2015 when we were in Ecuador. Artisanal fishermen from the local village cast a wide net around us in the bay we were surfing in using a small skiff. We got out of the water to watch about 20 local men haul in the gill net to catch the day's food. After much anticipation, the net yielded just half a bucket of puny fish, hardly enough to feed the fishermen, let alone their families.

International fishing policy is silencing our oceans. In 2018, global fishing subsidies totaled $35 billion.[312] Over 62 percent of that (or $22 billion) is considered by marine biologists as harmful subsidies that help increase intense industrial fishing capacity, often in distant foreign waters.[313] Just like many countries have agricultural policies that encourage crop overproduction that damages soils, industrial fishing subsidies deplete the natural capital of fish stocks. Instead of addressing overfishing, the root cause of fish depletion within their economic zones, many wealthy countries subsidize their fishing fleets so they can operate far from home in the poorly managed waters of developing nations or just outside them. The highest subsidizing nations in 2018 include China (the highest at 21 percent of global subsidies), Japan, Korea, the EU, and the US.[314]

Harmful subsidies include schemes that make diesel fuel cheaper for fishers and allow fishing fleets to travel much farther than is economical. Between 17 and 30 percent of these harmful subsidies are supporting fishing within foreign waters while 3–7 percent allow exploitation of the high seas.[315] It is a modern form of ecological colonialism: outsourcing the impacts of environmental overshoot to nations with fewer defenses.

**311** Charles Clover, *Rewilding the Sea* (Witness Books, 2022): xvii.

**312** U. Rashid Sumaila et al., "Updated Estimates and Analysis of Global Fisheries Subsidies," *Marine Policy* 109, 103695 (2019): sect. 3.1, https://doi.org/10.1016/j.marpol.2019.103695.

**313** Sumaila et al., "Updated Estimates," sect. 4.

**314** Sumaila et al., "Updated Estimates," sect. 3.4.

**315** Daniel J. Skerritt et al., "Mapping the Unjust Global Distribution of Harmful Fisheries Subsidies," *Marine Policy* 152, 105611 (2023): 5–6, https://doi.org/10.1016/j.marpol.2023.105611.

Just as the unsustainable destruction of megafauna on land forced people to start farming, our pillage of wildlife in the ocean is forcing people to farm fish. A parallel scenario of declining productivity ending in ecosystem collapse is inevitable. Farmed fish now make up half of global seafood consumption. Yet feeding many species of farmed fish, like salmon, depends on harvesting massive amounts of small-forage fish like krill, anchovies, and sardines. These are food sources for the ecologically important baleen whales and many other species.

Krill, now the most-harvested species in the Southwest Atlantic sector of the Southern Ocean, are under pressure. The Commission for the Conservation of Antarctic Marine Living Resources allows a catch limit for krill of 620,000 tonnes per year, but there is doubt, even within the organization, that even a tiny percentage of the krill harvest is sustainable for the other Antarctic animals like penguins that depend on it.[316]

To produce 1 kg of pricey farmed salmon, between 1.66 kg and 5.77 kg of small-sized fish species like anchovies are harvested to make fish pellets to feed them.[317] These oily fish could otherwise have fed people in the low-income countries where they are sourced. We are fast approaching a tipping point: a silent ocean devoid of playful dolphins, tuna, haddock, and other species that had survived ocean predation for millions of years.

I want my grandchildren to be able to take their grandchildren fishing and not fear powerful coastal storms driven by climate warming. I want future generations of artisanal fishers to be able to provide for their families. That future depends on holistic, regenerative fisheries management, and time is running out. It requires subsidies for industrial fishing to end now and marine protections to be enforced. Consumers need to understand what's at stake and make informed, holistic choices about what they buy and eat, and which policies they support.

The World Trade Organization has begun addressing this issue. The Agreement on Fisheries Subsidies was accepted into force on September 15, 2025.[318] It aims to redirect harmful subsidies toward sustainable practices. If countries comply, phasing out harmful industrial fishing subsidies could boost global fish stocks by 12.5 percent by 2050, adding 35 million tonnes of fish to the ocean's natural

**316** George M. Watters et al., "Long-Term Observations from Antarctica Demonstrate that Mismatched Scales of Fisheries Management and Predator-Prey Interaction Lead to Erroneous Conclusions About Precaution," *Scientific Reports* 10, no. 1, 10:2314 (2020): Results and Disc., https://doi.org/10.1038/s41598-020-59223-9.

**317** Spencer Roberts et al., "Feeding Global Aquaculture." *Science Advances*, 10, no. 42, (2024): table 1, DOI:10.1126/sciadv. adn9698

**318** World Trade Organization, "WTO Agreement on Fisheries Subsidies Enters Into Force," news release, September 15, 2025, https://www.wto.org/english/news_e/news25_e/fish_15sep25_e.htm.

capital, according to the PEW Charitable Trust.[319] The Pacific Ocean could see a 19 percent recovery in fish stocks. But don't count your sea turtles before they hatch. Competition for fish will be a fishbone of contention between countries in the coming years if stocks continue to be depleted as pollution and climate disruption impacts on marine life.

Merely fishing "sustainably" to maintain current severely degraded fish stocks keeps abundance low. Industrial fishing bycatch of dolphins, sharks, seabirds, and penguins is driving many species to extinction. Shark finning kills an estimated 80 million sharks per year![320] In comparison, only nine human deaths from shark attacks were confirmed in 2025.[321] Bottom trawling damages the sea floor mechanically, mirroring the way arable farming degrades soil when tillage is used. This destroys habitat and productivity. Ocean restoration should matter to everyone because, like land regeneration, it helps cool the planet and feed people and wildlife better.

It is time for individual countries to manage their coastal waters to restore abundance and for global policy to protect the high seas if we want fish to be on the menu for our grandchildren. Simply envisioning a teeming ocean with many whales spouting at once, bait balls shimmering everywhere, and dolphins and swordfish spinning in the air will not restore the ocean. Realizing the productivity potential of the ocean can help guide our actions but we must intentionally lobby for nature and push our governments to pass legislation to protect and restore oceans. We cannot rest if we want the ocean to have rest. We must make it happen because it is not the ocean that needs to change, it is people.

If fishing is reduced, how will those working in the industry earn a living? The fishermen of Cabo Pulmo, Mexico, found new jobs in eco-tourism as the fisheries they protected rebounded. Now the Cabo Pulmo Marine Park, which the local community lobbied their government to create 30 years ago, is one of the most popular diving destinations in the world. It brings in 8 million dollars in revenue to local communities.[322] The stewardship of natural resources by local people

---

**319** PEW Charitable Trust, "Ending Harmful Subsidies Could Increase the Amount of Fish in the Ocean, Research Shows," news release, June 7, 2021, https://www.pew.org/en/research-and-analysis/articles/2021/06/07/ending-harmful-subsidies-could-increase-amount-of-fish-in-the-ocean-research-shows.

**320** Boris Worm et al., "Global Catches, Exploitation Rates, and Rebuilding Options for Sharks," *Marine Policy* 40 (2013): Abstract, https://doi.org/10.1016/j.marpol.2012.12.034.

**321** Florida Museum, *Yearly Worldwide Shark Attack Summary: The ISAF 2025 Shark Attack Report*, (2026), https://www.floridamuseum.ufl.edu/shark-attacks/yearly-worldwide-summary/

**322** Octavio Aburto-oropeza et al., "Marine Prosperity Areas : A Framework for Aligning Ecological Restoration and Human Well-Being Using Area-Based Protections," *Fronteirs in Marine Science* 11, (2025): 10, https://doi.org/10.3389/ fmars.2024.1491483.

here has inspired a new concept of community-driven marine restoration called "Marine Prosperity Areas" that blends goals of marine restoration with human prosperity. The model takes a markedly holistic approach with "Community Engagement and Co-design," which brings together sometimes opposing groups of stakeholders to create a shared socio-economic, ecological, and cultural vision of where they want to go with the natural resource they jointly depend upon.[323]

The success of future Marine Prosperity Areas depends on government policies that provide targeted protection and support people in the fishing industry while fish stocks rebound. When fishers resume fishing in the Restoracene, humanity will be able to harvest far bigger catches from just the fish that make up the "interest" of a larger fisheries' natural-capital account. Fishing efforts (and jobs) in the meantime can be redirected instead to restorative seaweed and shellfish aquaculture that filters water, stores carbon and allows fishers to work closer to home. Consumers who want a future where they can take their grandchildren fishing don't need to wait for governments to solve the crisis. They can enjoy eating regenerative oysters, mussels, and seaweed instead of wild fish until the ocean can heal itself.

For fish lovers like me, giving up fish is not easy. Sushi and poke bowls are a favorite, healthy option for me when I am away from home. But today I was reminded why it may be worth taking ocean wildlife out of my diet. This morning, I paddled out for a surf at my local surf break and noticed two Hector's dolphins, a threatened species in New Zealand, breaking the water surface nearby. When such expert surfers appear in the lineup, I find myself under pressure to catch my waves so I don't make a fool of myself in front of them. I managed to make the next wave and glimpsed the two dolphins at my heels riding the wave their style—inside it.

As I paddled back out to the lineup, I saw them hurtling toward me inside a wave as I ducked under it. There is no doubt in my mind that they were having as much fun as I was as we surfed together. Their motivation could only have been to have fun, because there was no food for them in those waves. I know in my heart that this amazing species, and all the others that have spent millions of years evolving alongside ours, deserve to thrive on our collective home planet. Taking fish out of my diet until pristine ocean productivity is restored is the least I can do for them. The sooner more people take action toward this outcome, the sooner fish abundance will be restored.

---

**323** Aburto-oropeza et al., 6.

# Chapter 13

# Fire and Brimstone

Othering nature as if it were a mere resource to plunder instead of seeing ourselves as part of nature is no longer just unfortunate—it is dangerous. Southern California offers a prime example. There, many people spend hours in their cars on concrete roads, commuting to workplaces and schools with concrete floors. Aside from playing the odd soccer game, most people in urban areas rarely walk on soil or notice whether it soaks up rain or lets it run off until they are impacted by flooding. Most drivers pass by undeveloped land without realizing that these untended landscapes are often devoid of wildlife and are slowly deteriorating. This is a ticking time bomb.

In the open areas between towns and developments, grass plants are dying because grazing animals are no longer there to prune and fertilize them. Dead grass gives way to bare soil, which has been eroding for decades, as evidenced by the deep gullies etched into the landscape. Without the original complement of wild megafauna or even holistically managed livestock to keep woody vegetation in check, brush grows tall, building up fuel for wildfires and creating "ladders" that allow flames from grassfires to leap up into the trees. Tree canopies that once were spaced apart on pristine savanna are now connected, forming superhighways for hot, uncontrollable wildfire. Meanwhile, millions drive past these declining ecosystems, unaware of the dangers unfolding due to collective failure to manage land holistically.

## Stop Blaming Climate Change

The forest and shrubland that burned around Los Angeles in January 2025 revealed the need to cycle the carbon building up in these ecosystems biologically *before* it is released in to the air in costly infernos. The frequency of catastrophic wildfires doubled worldwide between 2003 and 2023.[324]

**324** Calum X. Cunningham et al., "Increasing Frequency and Intensity of the Most Extreme Wildfires on Earth," *Nature Ecology & Evolution* 8, no. 8 (2024): 1421, https://doi.org/10.1038/s41559-024-02452-2.

Much of the media blames climate change. While climate volatility can certainly heighten the risk of wildfires,[325] decades of poor land management has already made uncontrollable wildfires inevitable in so many places. Holistic land management to revegetate scarred land and restore the water cycle could save economies billions of dollars in wildfire losses and suppression costs.

## Southwest US

Many national parks in the American Southwest are missing the full complement of grazing animals that once prevented dead grass and dry shrubs from building up. For millions of years, megafauna were central to nutrient cycling—especially carbon—and helped protect living landscapes from fire. On the well-stocked Paleolithic savannas that once covered the American Southwest, browsers and grazers, such as giant ground sloth, mastodon, elk and bison, pruned grasses and shrubs, encouraging green regrowth. They paid carbon forward to smaller creatures like dung beetles, earthworms, and termites in the form of trampled branches, leaves, and manure. These organisms, in turn, passed the carbon to decomposers, like insects, bacteria, and fungi, that finished the cycle by breaking down organic matter into building blocks for humus, the carbon-based scaffolding of soil.

This process transferred carbon from flammable aboveground vegetation into humus-rich soil protected underground. This deep carbon sponge stored water and extended the green growing season, making fire even less likely. The presence of green vegetation for longer periods increased humidity because plants and trees can transpire moisture from the carbon sponge to the air, increasing rainfall and shortening dry spells.

Today, the land in some US state and federal parks where livestock are banned is spiraling toward desert because there are too few wild herbivores to fertilize and stimulate healthy, more fire-resistant grassland. I took a photo of oxidizing grass in California's Torrey Pines State Natural Reserve on a family visit in May 2019 (**Figure 13.1**). If I wanted to start a fire in my woodstove, this material would be the ideal tinder. Seed heads from the previous season's grasses lingered because they were never biologically recycled, and bare soil was visible under the woody brush. Everything was brown and tinder dry, even though it was the end of the wet season. Accumulated deadwood and leaves create a fire hazard, especially in California's Mediterranean climate. Add the fierce Santa Anna winds and you have a recipe for catastrophic inferno.

---

**325** Brendan Byrne et al., "Carbon Emissions from the 2023 Canadian Wildfires," *Nature* 633, no. 8031 (2024): 836–38, https:// doi.org/10.1038/s41586-024-07878-z.

**Figure 13.1. Oxidizing grass plants and dry woody shrub material at Torrey Pines State Natural Reserve, California, US**

Image by Siobhan Griffin (May 2019).

At Torrey Pines, the gray color of the dead grass tells us it is very slowly oxidizing into the air instead of being biologically cycled. In semiarid climates, when dead grass leaves do not have enough moisture to rot away, they decay chemically like this at a snail's pace. This blocks light from reaching the growing points of plants, blocking new green growth from forming and eventually killing the grass. A dead root system can no longer act as an anchor for the plant, and the wind or a hiker's footstep can shift the entire plant away, leaving behind bare, vulnerable soil. That's why so much ground at Torrey Pines and every other State Park I have hiked at in Southern California shows the signs of erosion that always results from bare soil.

Livestock could keep the grasslands healthy if managed in nature's image, but they are not allowed into Torrey Pines and other parks—even though these ecosystems evolved alongside abundant wild grazing animals that disappeared when humans arrived. What does this mean for nearby communities? Without animals to cycle the carbon biologically, the inevitable outcome is that the carbon in the vegetation will oxidize into the atmosphere all at once in an inferno.

## Hawaii

On the dry side of the Hawaiian Islands, including Maui, invasive grasses and brush have grown on former pineapple plantations and in between housing developments. There are too few grazing animals behaving naturally to keep this biomass under control. This neglectful land management created the conditions that led to the devastating Lahaina fire in August 2023.

The Lahaina Fire Incident Analysis Report issued by the Fire Safety Research Institute (FSRI) in September 2024 identified unmanaged vegetation on state and fallow private land upwind of Lahaina as a major predisposing factor.[326] After a similar fire in August 2018, three separate reports identified biomass buildup as a major risk, but little action followed. Instead, efforts were almost exclusively focused on fire suppression, not prevention.[327] This is a classic case of people failing to address the root cause of the problem.

Managing vegetation on state land had proved difficult. Herbicides were used to kill the fire-resistant grasses, but they didn't remove the material, and workers had to remove it manually, which was so inefficient little was achieved. Firebreaks made with backfires and bulldozers caused soil erosion because of the bare soil they created.

The FSRI report recommended that returning the land to native ecosystem cover or agriculture would help reduce fuel buildup and grow food. They went as far as recommending targeted grazing to reduce biomass of non-native grasses.[328] Holistic planned grazing would not only recycle vegetation, but it would also keep it under control long-term, while creating jobs and producing food. The costs would be far lower than with mechanical or chemical control methods.

## Grazing to Prevent Catastrophic Fire

The preventive approach of herding creates jobs, produces food, and can save lives. Goats, donkeys, and sheep can graze under power lines on rough terrain to suppress fuel where mowing is impractical, helping utility companies control vegetation without using poisonous herbicides. Holistic management educator Cole Bush herds mobs of 400 goats and sheep to tackle overgrown, oxidizing grass and brush that is building up in California's hills.

---

**326** Steve Alkonis and Derek Kerber, *Lahaina Fire Incident Analysis Report* (FSRI, 2024), 59–62, https://doi.org/10.60752/102376.26858962.

**327** Alkonis and Kerber, *Lahaina Fire Incident Analysis Report*, 63.

**328** Kerber and Alkonis, *Lahaina Fire Incident Analysis Report*, 66.

The Los Angeles County Sheriff's Department teamed up with her company, Shepherdess Land & Livestock Co., in 2023 to train prison inmates in herding and holistic land management. The inmates get valuable hands-on experience using modern livestock tools like solar-powered fencing. When they return to society, they will have valuable skills for preventing wildfires and ecological stewardship. Teaching young inmates about ecology is a win-win, especially in a country like the US that has more prisoners than farmers.

## Nature's Design for Fire Resilience

Northern California's oak savanna ecosystem was intact 25,000 years ago, hosting groves of trees interspersed with grasslands that were kept green and productive much of the year by grazing megafauna like mastodons, bison, horses, and elk. When fires occurred, they burned low and cool and did not burn down mighty oaks and redwood trees. These open landscapes were shaped by grazing animals and were more fire-resilient because of them.

When I used to hike and ski in forested areas of New York and Vermont, I imagined the dark, closed-canopy forest was the region's pristine condition. But, when native prairie species like ground-nesting bobolinks appeared on our New York State farm and began to thrive in large numbers, I realized that I was witnessing a glimpse of the pristine ecosystem that had once thrived in the Eastern US, and it wasn't dense forest.

It turns out megafauna had participated in keeping productive American ecosystems open throughout much of Canada, the US, and Mexico and had brought their prairie friends with them. Early human explorers to the Americas would have encountered giant bison *Bison latifrons*, with horns spanning over two meters in breadth. These large ruminants, once found throughout North America as far down as Southern California and Mexico, disappeared around 20,000 years ago as brave human hunters exploited North American megafauna.

In temperate, humid areas like the Northeast US, unused land has unnaturally reverted to forest because abundant roaming herbivores are gone. Fossil records show that horse, mammoth, and ground sloth once roamed New York State, meaning the forest on all but the steepest land was once more open. Around 116,000 years ago, during the warm interglacial period before the one we are in now, mosaic landscapes of open grassy areas and open woodlands maintained by abundant megaherbivores covered over 50 percent of Europe.[329] *Homo sapiens* had yet to arrive in either Europe or the Americas at this time, which suggests that open woodlands, not dense forests, were the default in eastern American temperate lowlands, too.

## The Tool of Rest Along the Brittleness Gradient

One of Allan Savory's most insightful observations was that ecosystems react differently to the same management tool depending on rainfall reliability. He created the "brittleness scale" to describe it. Understanding where your ecosystem falls on the scale is very important for farmers and non-farmers alike because it helps people know which tools will be most effective to avoid forest fires and manage grazing. Land is rated from 1 to 10, with 1 being humid rainforest and 10 being dry or seasonally dry desert. Environments will respond in different ways to the five land management tools available to humans: fire, rest, technology, human creativity, and grazing. We must match the right tools to the landscape's brittleness if we want to move closer to the future landscape we want.

## A Tale of Two Grasslands: Reliable Rain

Let's look at the differences in outcomes when the land management tool of rest is applied to non-brittle and brittle ecosystems. Rain is reliably distributed throughout the year in central New York and Vermont, where I farmed, placing the Northeast US at the non-brittle end of the scale. The hills around Raindance Farm get 700 to 1,000 mm of precipitation per year, and rain or snow usually falls in every month of the year.

Forest had overtaken the land in the wake of megafaunal extinctions 13,000 years earlier, although animal impact from grazers and management by indigenous people probably kept it somewhat open. In the 1800s, trees in central New England and New York were cleared for grazing. By the mid-19th century, 80 percent of the forest across central New England had been converted to farmland.[330]

---

**329** Elena Pearce et al., "Substantial Light Woodland and Open Vegetation Characterized the Temperate Forest Biome before Homo sapiens," *Science Advances* 9, no. 45 (2023): 2–4, https://doi.org/10.1126/sciadv.adi9135.

**330** David R. Foster, "Land-Use History (1730–1990) and Vegetation Dynamics in Central New England, USA," *Journal of Ecology* 80, no. 4 (1992): 753, https://doi.org/10.2307/2260864.

After World War II, wool prices collapsed, so farmers stopped raising sheep in the Northeastern US because it was not as profitable as dairy farming. This gave the steeper hills that had been suited to sheep, but not cattle, a rest from grazing. Also, the Agricultural Act of 1956 and its Soil Bank Program paid farmers to "rest" cropland on 12 million sloped and vulnerable hectares across the country to reduce soil erosion.[331]

Since sheep no longer dotted the hills to prevent woody plants from growing, trees gradually shaded out the grasses. Over time, second-growth forests covered the steep slopes. Today, over 80 percent of the Northeast US is covered in native regenerated forest.[332] These forests, full of second-growth maple, oak, pine, ash, and hemlock, create the beautiful tapestry of autumn colors that draws thousands of tourists to the region every October. Hikers wandering in the woods in New England may even notice the remnants of stone walls that were built to keep in the sheep that used to roam the hills. These landscapes are the legacy of applying the tool of rest from grazing in a non-brittle environment, which results in succession to closed-canopy forest only when sufficient wild megafauna are not present.

## A Tale of Two Grasslands: Seasonal Rainfall

The semiarid Serengeti region of East Africa receives a similar amount of annual rainfall to Raindance Farm: about 1,000 mm. But the landscape on the Serengeti will degrade if the tool of rest is applied because there is a six-month dry season, and it experiences higher temperatures. Savory considers these ecosystems to be "brittle"—leaves left on plants after the growing season are a crispy liability. If grazing animals are taken away or not managed properly to trample or remove the dead leaves, the grass plants shade themselves to death just as we saw at Torrey Pines State Park.

If too few herbivores graze a semiarid landscape, they tend to selectively overgraze the most palatable plants to death. The unpalatable grass plants that do not get grazed or trampled die eventually, too, and bare soil is exposed to wind and water erosion. Over time, the region turns into desert, as Savory has been documenting in both Africa and the US Southwest over the past 40 years.

---

**331** Zachary Cain and Stephen Lovejoy, "History and Outlook for Farm Bill Conservation Programs," *Choices* 19, no. 4 (2004): 38, https://www.choicesmagazine.org/2004-4/policy/2004-4-09.htm.

**332** C. W. Woodall et al., "Monitoring Network Confirms Land Use Change Is a Substantial Component of the Forest Carbon Sink in the Eastern United States," *Scientific Reports* 5, no. 1 (2015): 2, https://doi.org/10.1038/srep17028.

## Grazing Can Prevent Wildfire and Sequester More Carbon

The role of large grazing animals for fire control is made clear in a chapter from East Africa's history. From the late 1800s to the 1960s, a rinderpest outbreak decimated wildebeest, buffalo, and other grazer populations. Without these animals to cycle biomass, wildfires burned up to 80 percent of all plant growth—including trees.

As wildebeest numbers recovered, they began converting grass biomass into fertilizer, which dung beetles and other soil life could pull underground. Their hooves trampled flammable woody brush, making it easier for termites to process. The carbon cycle was restored and assisted a healthier water cycle. The savannas returned to their mosaic form of grasses punctuated by scattered trees. The Serengeti had shifted from being a net carbon source to a net sink with the return of the wildebeest. Wildebeest activity reduced fire, increased tree density, and increased carbon sequestration by removing 40–70 tonnes of carbon per $km^2$ from the atmosphere. But the Serengeti will only continue as an efficient carbon sink in the short term with the help of at least one million wildebeest.[333]

It is logical, then, to consider introducing holistically managed livestock onto overgrown semiarid landscapes in Southern California to help prevent wildfires. As of January 2024, there were over five million cattle in California.[334] Most are not being managed to use their nutrient-cycling superpowers to enhance ecosystem restoration. Although 1.7 million of these cows are dairy cows that are unsuitable for rough terrain, with purposeful breeding, even they could be bred to be fit-for-purpose land healers. Other animals, such as donkeys and camels, are already adapted to help regenerate degraded deserts. Properly managed, they too can help restore wildlife habitat, water cycles, and carbon sinks across fire-prone regions.

## Preventing Wildfire in Australia

Donkeys are particularly good at chewing down tough dried grasses on hilly, tough terrain during drought. In Western Australia's Kimberley region, rancher Chris Henggeler is managing wild donkeys and feral cattle to reduce the severity of wildfires. His family ranch, Kachana Station, uses salt blocks to attract donkeys

**333** Ricardo M. Holdo et al., "A Disease-Mediated Trophic Cascade in the Serengeti and Its Implications for Ecosystem C," *PLOS Biology* 7, no. 9, e1000210, (2009): Disc., para. 6, https://doi.org/10.1371/journal.pbio.1000210.

**334** National Agricultural Statistics Service, "2024 State Agriculture Overview: California," Accessed August 7, 2025. https:// www.nass.usda.gov/Quick_Stats/Ag_Overview/stateOverview.php?state=CALIFORNIA#:~:text=USDA/NASS%20 2024%20 State%20Agriculture,40%2C283%2C000%2C000.

to rocky hills that are too steep for cattle, encouraging them to graze down dead, flammable grass and brush. Once the area is grazed, they move the attractants to allow for plant rest and recovery. The result: thicker grass cover and improved soil retention.[335]

Henggeler is reversing the forces of desertification on his land, even while he still sees surrounding landscapes continue to degrade under poorly managed grazing. He has flown over the region for decades and sees the contrast firsthand. The Australian government recently ordered him to slaughter his donkeys because of the damage many unmanaged donkeys inflict by overgrazing many other areas in Australia. Chris insists on defending his small donkey herd and was even prepared to go to jail for refusing to slaughter them because when he had destocked the ranch previously, ungrazed grass built up and fueled wildfires.

The unmanaged donkeys and feral cattle outside Kachana Station roam without predators, so they don't bunch up or move on like herding animals do. Instead, they will continue to eat the most nutritious plants repeatedly until they die, leaving only tough or poisonous plants behind. Set-stocked cattle or sheep do the same. This is why the landscape outside his farm continues to desertify. Chris now calls for more "land doctors"—ecologically trained professionals paid to heal ecosystems. He believes restoring nature's green soil armor benefits everyone by cooling the planet and restoring water cycles.

## The Disappearance of Megafauna in Canada

The lessons from East Africa also apply to former savannas in cooler, wetter regions. Alberta, Canada, is less brittle than the Serengeti, even though it gets only about half as much rainfall. This is because Alberta has cooler temperatures and a more even year-round rain distribution. After the glaciers retreated around 15,000 years ago, megafauna like giant ground sloth, horse, mammoth and bison fertilized the recovering prairie west of the Laurentide Ice Sheet. Northern Alberta once supported vibrant grasslands that stretched as far north as the Yukon. As humans first arrived, they reduced the impact of giant ecosystem engineers, allowing forests to spread and fuel loads to build up. Then between 1750 and 1880, European settlers nearly eradicated plains bison (*Bison bison*) and wood bison

**335** Monique Ross and Ben Cheshire, "Meet the Defiant Grazier Using Feral Donkeys to Regenerate His Land at Kachana Station in an 'Illegal' Experiment," ABC News, November 3, 2024, https://www.abc.net.au/news/2024-11-04/grazier-using-wild-donkeys-to-regenerate-land-in-legal-battle/104445766?future=true&.

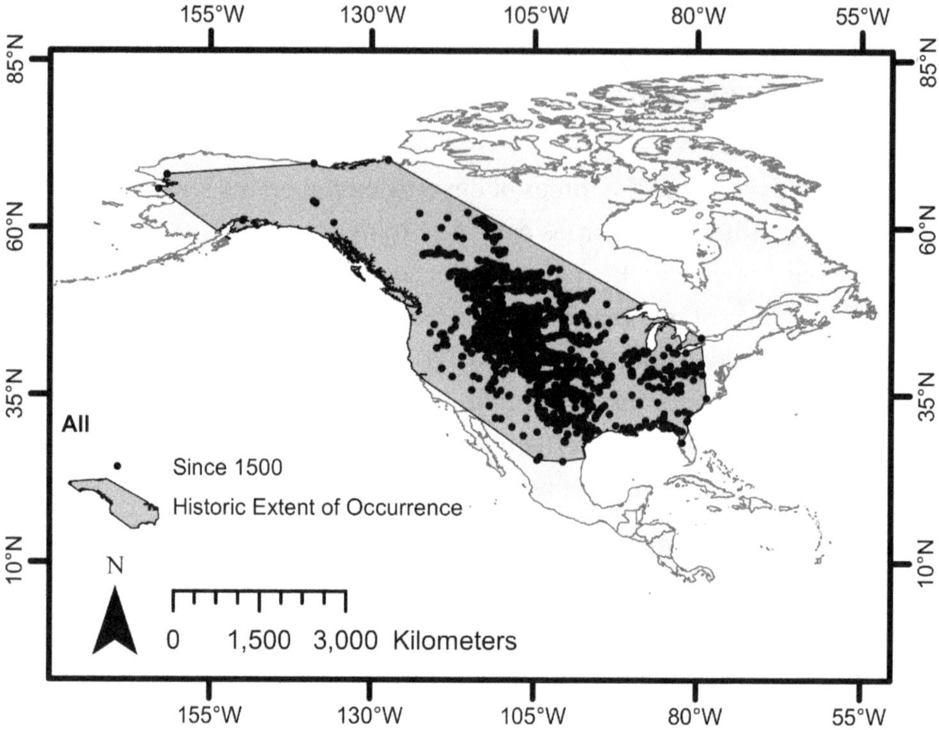


**Figure 13.2 Bison observations since 1500 A.D. integrating 3 disciplines of historical ecology, archaeology, and paleontology to establish the historical extent of occurrence (dark gray polygon with black border).**

Source: Martin et al., (2022).

(*Bison athabascae*) that once numbered in the millions in a historical blink of an eye.[336] Resting land from grazing increases tree density when enough moisture falls regularly, and the removal of grazing and browsing megafauna has increased fire severity and frequency in many regions of the world as a result.[337] Without sufficient numbers of grazers to create firebreaks of short grasses, former bison and mammoth range in Canada became much more prone to uncontrolled wildfires ignited by lightning or humans. The historical extent of bison since 1550 based on integrated evidence is outlined in **Figure 13.2**. Bison once stretched from central Alaska down to Mexico and from California in the West, to as far East as Florida.

**336** Jonathan James Farr and Clifford A. White, "Buffalo on the Edge: Factors Affecting Historical Distribution and Restoration of Bison Bison in the Western Cordillera, North America," *Diversity* 14, no. 11: 937, (2022): sect 1, https://doi.org/10.3390/ d14110937.

**337** Christopher N. Johnson et al., "Can Trophic Rewilding Reduce the Impact of Fire in a More Flammable World?," *Philosophical Transactions of the Royal Society B: Biological Sciences* 373, no. 1761 (2018): sect. 3, para. 6, https://doi.org/10.1098/ rstb.2017.0443.

This slaughter of bison herds by European settlers allowed more flammable softwood forests to encroach on former prairie and open forest ecosystems. Today in Alberta, outside of the cropping areas of the wheat belt where forest is kept at bay with farming, aspen, spruce, and pine forest has replaced prairie that once was prevalent near towns named after former grassland landscapes such as "Buffalo Head Prairie" and "Grande Prairie."

On a single day in October 2023, there were 133 fires in Alberta and 54 in Saskatchewan. That year's wildfire season in Canada emitted 647 million tonnes of $CO_2$ back into the atmosphere. This was more than the total greenhouse gas emissions of most of the larger countries in the world except the US, China, and India.[338] Canadian forestry emitted more $CO_2$ equivalent pollution than the country's energy sector, which includes dirty tar sands exploitation. The areas with the highest intensity of fires sit squarely within the footprint of former bison range.

## To Tree or Not to Tree

Holistically managed grasslands can be a much safer place to sequester carbon than forest over decades to centuries in drought- and fire-prone regions like California because two-thirds of the carbon-rich biomass is below ground, protected from fire.[339] Scientists are beginning to recognize that creating carbon models to inform climate mitigation without including the impacts large herbivores have on carbon and nutrient cycling is flawed.[340]

Some forests, like the Amazon Rainforest, are vital for creating rain and cooling the planet, but plantation monocultures of pine, fir, or spruce don't offer long-term carbon storage. These fast-growing but weak timbers are often used to make products that do not last long like toilet paper, cardboard, and pallets. Once harvested, the carbon in tree biomass returns to the atmosphere through decay or combustion, usually within 30–50 years even if it is used for framing timber. Unlike forests, grasslands increase soil carbon for storage over millennia.[341]

**338** Byrne et al., "Carbon Emissions," 835.

**339** Pawlok Dass et al., "Grasslands May Be More Reliable Carbon Sinks than Forests in California," *Environmental Research Letters* 13, no. 7, (2018): 7, https://doi.org/10.1088/1748-9326/aacb39.

**340** Matteo Rizzuto et al., "Rewiring the Carbon Cycle: A Theoretical Framework for Animal-Driven Ecosystem Carbon Sequestration," *Journal of Geophysical Research: Biogeosciences* 129, no. 4 (2024): 14, https://doi.org/10.1029/2024JG008026.

**341** Gregory J. Retallack, "Global Cooling by Grassland Soils of the Geological Past and Near Future," *Annual Review of Earth and Planetary Sciences* 41, no. 1 (2013): 70, https://doi.org/10.1146/annurev-earth-050212-124001.Histosols

In New Zealand, planting pine monocultures on former grasslands has resulted in losses of soil carbon.[342] As the soil under plantation forestry loses organic matter, it becomes less effective at absorbing large rainfall events. In 2018, storms caused horrible erosion from forestry land, and thousands of tonnes of soil and debris were dumped into the sea at Tologa Bay, New Zealand. Similarly, the 2021 floods in Princeton, Canada, occurred in a region that has been heavily logged in recent decades. These monocultured landscapes lack the root structures to absorb rainfall or withstand erosion. Catastrophic flood events have accelerated since then as a warmer climate is more capable of dumping severe rain.

Consideration of the ability of grasslands to sequester carbon in soil over the long term must be part of our climate mitigation conversation. A 2022 study asked whether large herbivores could enhance long-term carbon persistence. The researchers concluded, "We need to shift our focus away from building immediate carbon stocks to also include carbon persistence, as increased persistence may be needed to secure long-term carbon sequestration in a world of more frequent and severe perturbations."[343] Radiocarbon data show that carbon in the top meter of soil is on average 4,830 years old! Topsoil carbon (0–30 cm) is 1,390 years old, and in subsoils (30–100 cm), carbon averages 8,280 years.[344] The more we understand the symbiotic relationships between animals and plants that make ecosystems more productive, the better we can leverage nature's templates to support more abundance and help cool the planet.

That said, forests products can provide food and useful inputs for manufacturing that otherwise would be made from fossil fuel-derived raw ingredients. Like farming, agroecosystems inspired by nature can be leveraged in forestry to enhance biodiversity. For instance, silvopasture incorporates wood, fruit, and nut production into a mosaic with grassland production of livestock.

---

**342** Louis A. Schipper et al., "A Review of Soil Carbon Change in New Zealand's Grazed Grasslands," *New Zealand Journal of Agricultural Research* 60, no. 2 (2017): 98, https://doi.org/10.1080/00288233.2017.1284134.

**343** Jeppe A. Kristensen et al., "Can Large Herbivores Enhance Ecosystem Carbon Persistence?," *Trends in Ecology & Evolution* 37, no. 2 (2022): 125–26, https://doi.org/10.1016/j.tree.2021.09.006

**344** Zheng Shi et al., "The Age Distribution of Global Soil Carbon Inferred from Radiocarbon Measurements," *Nature Geoscience* 13, no. 8 (2020): 557, https://doi.org/10.1038/s41561-020-0596-z.

Greg and Rachel Hart, holistic management practitioners at Mangarara Station, New Zealand, have planted thousands of trees in the past 20-plus years, some with the help of Air New Zealand employees. A tapestry of apple, peach, cherry, plum, nuts, hardwoods, and softwoods line up across their pastures in intentional rows designed to make cattle movements easy. When I co-led a grazing management workshop for farmers at the Harts' eco-lodge in autumn 2025, the people attending were able to graze on the edible landscape right beside the herd of angus beef cattle.

The beating heart of the farm, Horseshoe Lake, was filled with many species of noisy ducks and black swans while insects buzzed about. When I went for a swim around the lake, I swam away from the herds of swans so I would not disturb them. When I paused to tread water and take a breather, I was surprised to see what was behind me. Hundreds of swans had followed me, curious to know what clumsy aquatic creature had joined them. The farm experience was like a visit to the Garden of Eden described in holy books, a reflection of the Restoracene in my mind.

# Chapter 14
# The Restoracene

Humanity stands at a crossroads. By robbing present generations of the wonder of amazing creatures like *Diprotodon optatum*, moa, mammoth, and mastodon, our ancestors changed the face of the earth. What wondrous creatures are we stealing from our children and grandchildren today? Many species have gone extinct under the watch of my generation around the planet in the last 60 years, including the Japanese otter (*Lutra nippon*), the Japanese sea lion (*Zalophus japonicus*), the Australian lesser bilby (*Macrotis leucura*), the Mississippi Valley wolf (*Canis rufus*), the Caspian tiger (*Panthera tigris tigris*), the Bali and Javan tiger (*Panthera tigris sondaica*), the Iberian ibex (*Capra pyrenaica*), the African western black rhinoceros (*Diceros bicornis longipes*), the Barbary lion (*Panthera leo leo*), and many smaller, less charismatic species. What will be the consequences of continued species loss for life on our planet?

## What Kind of Future Do You Really Want?

People know the importance of rainforests for producing oxygen, rain, and countless ecosystem services, but the rate of deforestation has accelerated against the will of most of our planet's inhabitants. Virtually all nonhuman life is declining, even though every earthling desperately needs the complex web of life that underpins our economies. Why do we behave in a way that thoughtlessly wastes the resources future generations will need? Do we really need so much "stuff" and the pollution associated with the production and disposal of so many consumer products? Relentless consumerism is using up resources for future generations at breakneck speed. Modern people did not ask for the world we now live in—it was sold to us by marketers. This is not the secure, abundant future most of us want for our descendants. It's time to say: this isn't what we ordered.

## Becoming a Good Ancestor

When we make decisions to solve individual problems, or to satisfy a want or need, we risk creating unintended consequences for future generations. For instance, if we buy a polypropylene workout shirt that is on sale, we think we have gained something useful—but at what cost? What pollution occurred during its production, and where will it go at the end of its life? Could this fossil-fuel-based product end up polluting the ocean for hundreds of years?

Taking actions, such as what you buy, within a holistic context that you create, solves this tricky problem. The holistic context is the guiding star that has been missing in the history of human impact on the planet. Without a holistic context to guide our consumer decisions, marketers will continue to woo us down a path that is destroying our inheritance of natural resources.

## On the Road to a Future We Never Asked for

In the paper "World Scientists' Warning: The Behavioral Crisis Driving Ecological Overshoot", scientists explain the futility of approaching symptoms of our multiple global crises that include species extinction, water degradation, and climate change, with one-at-a-time problem-solving approach using technology. It is human behavior that needs to change. After all, the core drivers of anthropogenic ecological overshoot are human behaviors and culture around consumption and population growth.[345]

Over the past 30 years of ramping up renewable energy, shifting to electric transportation, and producing 12 million electric cars to address climate change, our carbon pollution has only risen. Why? Because we haven't addressed the root causes of climate disruption. During that same period, the human population increased by 40 percent, far offsetting any climate gains from energy efficiency.[346]

Technology has been ineffective at solving multiple global crises because it targets symptoms, not causes. For instance, the US Department of Energy estimates that achieving carbon neutrality by 2050 using carbon-capture technology would take between 20,000 and 96,000 miles (32,000–155,000 km) of new pipelines to transport captured $CO_2$ from power plants.[347] The length of

345 Joseph J. Merz et al., "World Scientists' Warning: The Behavioural Crisis Driving Ecological Overshoot," *Science Progress* 106, no. 3 (2023): fig. 1, https://doi.org/10.1177/00368504231201372.

346 Aalok Ranjan, "Population Effects of Increase in World Energy Use and $CO_2$ Emissions: 1990–2019," *Journal of Population and Sustainability*, vol 5, (2020): 17, DOI: 10.3197/jps.2020.5.1.87.

347 E. Larson et al., *Net-Zero America: Potential Pathways, Infrastructure, and Impacts: Final Report Summary* (Princeton University, 2021), https://netzeroamerica.princeton.edu/img/Princeton%20NZA%20FINAL%20REPORT%20SUMMARY%20 (29Oct2021).pdf

this pipe could encircle the Earth four times and would require massive material inputs and fossil fuels to produce. In other words, it would create new unintended consequences. This reductionist approach won't solve crises such as species extinction, inequality, or ecological overshoot.

Often the energy "solutions" offered to us keep us trapped in our consumer roles. Electric cars are a prime example. The people promoting, manufacturing, and marketing electric cars for profit are happy to ignore what societies really need, which is a redesigning of cities and towns replete with all the efficient public transportation infrastructure we need to get where we need to go. Instead of having good choices around transportation, most citizens in the US are trapped into owning expensive cars to get around. Ecological overshoot results from people consuming natural resources faster than ecosystems can regenerate them and polluting more than the Earth can absorb. As scientists point out, this overshoot is driven by overconsumption, waste, and population growth—root causes that should come as no surprise to anyone.[348]

Our business-as-usual problem-solving habits are not fit for purpose. We must consider people and the planet as well in every economic decision. Universal education in holistic management could free humanity from self-destructive systems by giving us a way to choose a better future. But we face systemic obstacles on the path to the peaceful, environmentally sound future most people say they want.

## What Stands in the Way of a Better Future?

Scientists who are working in the limits to growth space identified one reason we're hurtling toward a self-destructive future: "In essence, power dynamics in society underlie the manipulation of needs, wants and desires." [348] These power dynamics can take many forms—relentless marketing by deep-pocketed corporations to encourage the consumption of products we don't need, corruption that facilitates greed, or pronatal pressure from religious groups. Power dynamics can be as blatant as propaganda coming from power-hungry politicians protecting their own interests.

**348** Merz et al., "World Scientists' Warning," 4.

## Power Dynamics

Our current global economic system, which enriches a small percentage of humans, requires relentless economic growth at the expense of your environment. Gross domestic product (GDP) is a measure of consumption. If countries continue to expand their GDP without limits, we will inevitably use up all the natural capital on the planet, leading to ecological collapse and nothing to eat. Still, powerful corporations fan the flames of consumerism with relentless marketing of nonessential products. Overexploitation of the natural resources future generations will need, including fossil fuels and topsoil, is rapidly converting your natural capital into unimaginable wealth for the elite. Natural capital is the birthright of all citizens of our planet, present and future, and should be stewarded to benefit everyone, not liquidated to enrich a few today.

Oxfam International recently revealed that the world's 12 richest people now hold more wealth than the poorest half of humanity (four billion people). Even more worrisome is the political power this wealth is buying. A billionaire today is 4,000 times more likely than the rest of us to gain a political office.[349] Many people tolerate injustices like this because we have been told stories to justify extreme greed. Trickle-down economics comes to mind even though I don't know anyone who can live well on a trickle. And why should we have to? All people should be paid fairly for the work they do. To give you an idea of what the "trickle" looks like in perspective: For every dollar someone in the bottom 90 percent wealth bracket earned between 2020 and 2023, a billionaire earned $1.7 million.[350]

The super-rich also benefit disproportionately from the degradation of your life-giving ecosystems through their investments. The investments of each of the richest 50 billionaires are responsible for almost 400,000 years of consumption emissions by the average person, or 2.6 million years of consumption emissions by someone in the poorest 50 percent of the world.[351] The consequences of exploitive industrialization are sprinkled among the less fortunate majority, who get left holding the bag for cleanup and have to pay the health costs of pollution.

With wealth comes power and influence—especially over the media, which the elite control with the ulterior motive of propping up the system that enriches

**349** Oxfam, "Billionaire wealth jumps three times faster in 2025 to highest peak ever, sparking dangerous political inequality says Oxfam," (Oxford: *Oxfam GB*, January 2026): 9, https://www.oxfam.org.uk/mc/ax2v6v/.

**350** OXFAM, "Survival of the Richest: How We Must Tax the Super-Rich Now to Fight Inequality," (Oxford: *Oxfam GB*, March 2023): 8, https://oxfamilibrary.openrepository.com/bitstream/handle/10546/621477/bp-survival-of-the-richest-160123-en.pdf

**351** Oxfam, "Carbon inequality kills: Why curbing the excessive emissions of an elite few can create a sustainable planet for all," policy and practice briefing paper, Oxford: *Oxfam GB*, (2024):12, DOI: 10.21201/2024.000039

them. Together, they are writing an inequitable future script for humanity that will be played out on a barren and polluted landscape. But you can take back autonomy over your destiny if you stop willingly playing their game and let a holistic context that you create guide your actions. The continuing functionality of the environment must be built into the holistic contexts and mission statements of businesses, as it underpins all economies, now and in the future. This could mean creating a sensible future context where your business stops seeking maximum growth for the benefit of investors and instead offers a fair investor dividend and meaningful, well-paid jobs. For consumers, it might mean buying products from employee-owned companies that have these values. This will prioritize sustainability over relentless growth.

It is time to put an end to profits made by large corporates at the expense of people and the planet. Corporations that destroy your natural capital to enrich their shareholders can change and become entities that provide good jobs while moving humanity closer to the Restoracene. Yvon Chouinard, owner of the outerwear company Patagonia, made a paradigm-smashing move in 2022 by putting company ownership into the Patagonia Purpose Trust, with a mission to benefit the environment.[352] Novel economic hacks like this will emerge in the Restoracene once enough people evolve in their decision-making to include holistic frameworks. But a level economic playing field must be provided by governments for new business models and economies to emerge. This requires enforcement of anti-trust laws and new legislation to outlaw lobbying.

One powerful feature of the holistic context is that it invites you to imagine yourself living within it. This gives you the chance to define yourself not as a worker/consumer but as a human being. You define the important networks of people you socially connect with in your realm of agency and describe how you want those connections to feel. For example, you may want to be seen as honest and curious. As a business owner, you might want to be known as a fair employer or a reliable supplier. Going forward, this will shape all your actions and decisions as you check them against your holistic context to confirm you're moving closer to the future you want. You can choose to be a caretaker of soul-healing, life-supporting natural spaces that future generations will need too. You can be what all humans were born to be: symbionts in both human societies and living ecosystems. [353] Asking holistic testing questions will keep humans empowered over our own destinies.

---

**352** Oxfam, "Carbon Billionaires, " 12.

**353** Savory and Butterfield, *Holistic Management*, 264.

## Oligarchs Block the Path to the Future You Want

Corporations controlled by the super-rich are demanding that we, the people, cede to them the legal right to liquidate our nation's natural resources. For instance, fossil fuel companies are racing to export American liquefied natural gas (LNG). Reuters reported that the US became the first country to export 10 million tonnes of LNG in single month in October 2025.[354] The trade in LNG enriches wealthy shareholders disproportionately while depriving our grandchildren of a resource that will be highly valuable for their economies in the future. North America can be a cold place, and our grandchildren will need energy to stay warm in the winter and to have air conditioning in warmer summers. What expensive choices will we leave to them? Even the manufacture of renewable energy infrastructure requires fossil fuels.

Corporations also demand the right to pollute with impunity in the name of "freedom" from commonsense regulation. The United States Environmental Protection Agency (EPA) reported that on February 28, 2025, Congress passed a bill repealing a fee on the very worst polluters of methane in the oil and gas industry.[355] The proposed fee had been part of the 2022 Inflation Reduction Act and was designed to reduce air pollution and conserve methane—a potent greenhouse gas—rather than let producers waste it by releasing it into the air as a byproduct of oil and gas production.

With this law, money collected in fines from the companies that wasted the most natural gas during their fossil-fuel extraction activities was going to help fossil fuel companies deploy technology to stop gas leaks, help states monitor methane leaks, and plug abandoned polluting oil and gas wells. Reducing the wastage of methane during oil and gas production could have conserved more fuel for our grandchildren to use and reduced climate-warming pollution from oil and gas extraction. Now, the costs will be passed along to the public who will bear the costs of both pollution and the climate instability fossil fuel companies are accelerating far into the future. Why is legislation like this, that is meant to protect the health and prosperity of all citizens, so hard to enact?

**354** Curtis Williams, "US Becomes First Country to Export 10 Million Tonnes of LNG in Single Month," Reuters, November 5, 2025, https://www.reuters.com/business/energy/us-becomes-first-country-export-10-million-tonnes-lng-single-month-2025-11-03/.

**355** EPA, "Waste Emissions Charge," accessed January 11, 2026, https://www.epa.gov/inflation-reduction-act/waste-emissions-charge.

## Legalized Corruption

The influence of corporations and their lobbying groups has become so pervasive in US politics that the average American's voice is nearly drowned out. Even when most citizens may oppose legislation, influential business-oriented lobbying groups supporting it often get their way. A 2014 statistical analysis of legislation passed in the US found, "When the preferences of economic elites and the stands of organized interest groups (lobbyists) are controlled for, the preferences of the average American appear to have only a minuscule, near-zero, statistically insignificant impact upon public policy."[356] Meanwhile, policies backed by lobbying groups have a 40 percent chance of being enacted. The resulting problem is loss of autonomy over your future. Even if most citizens vote for a future they believe in, special interest lobbies stand in their way. Doesn't this make lobbying a legalized form of corruption? The problem of lobbying is a bipartisan issue.

Speaking of the two-party political system, it is worth noting that this system is clearly not working for the benefit of the people in the US. When a party gets elected, they spend four years trying to stay in power instead of doing the work their electorate really needs. Surely a better political system can evolve when people acknowledge the need for change.

## The Curtain of Confusion

Lack of transparency is the playbook often used by political cronies, corporations, and media barons to manipulate the media. My struggle around the implications of global climate disruption became an emotional battle as I struggled for the past 40 years with climate change denial perpetrated by fossil fuel corporations with ulterior motives. I lived and worked at the sea for the Town of Southampton, New York, as a lifeguard in the early 1980s. I noticed seawater from the bay flooding roadways at high tide on fine sunny days in places it never had before. The dunes at my favorite surfing beach all washed away. I knew the sea was rising. I could see it with my own eyes. But the truth is inconvenient for fossil fuel companies and other high-consumption industries.

**356** Martin Gilens and Benjamin I. Page, "Testing Theories of American Politics: Elites, Interest Groups, and Average Citizens," *Perspectives on Politics* 12, no. 3 (2014): 575, https://doi.org/10.1017/S1537592714001595.

Back in 2011, a joint report by researchers from Cornell University, Columbia University, and the City University of New York had warned that accelerating rates of ice melt from the Greenland and the West Antarctic ice sheets could result in sea level rise on the order of 37–55 in (55–140 cm) by the 2080s.[357] But I never heard this report covered in the mainstream news and it didn't enter public debate. The media who claimed to be covering the news spent far more time on entertainment than they did shedding light on emerging science around climate disruption. Since then, greenhouse gas and sea levels have only risen. Decades of inadequate coverage by the media around the climate crisis have cost us precious time to realize we need to stop polluting the air we breathe yesterday, not tomorrow. The elite controlling the media are the most wasteful and polluting people on the planet. Only if we demand transparency, seek the truth, and come together to fight for the future we truly want and need, can the Restoracene be possible.

**357** New York Academy of Sciences, "Responding to Climate Change in New York State: The ClimAID Integrated Assessment for Effective Climate Change Adaptation in New York State," *Annals of the New York Academy of Sciences* 1244, no. 1 (2011): 39, https://doi.org/10.1111/j.1749-6632.2011.06331.x.

# Chapter 15

# Extinction Redemption

Powerful interests are writing a script that continues the liquidation of your grandchildren's natural capital to enrich themselves. Nowhere is our neglect to shape our grandchildren's future more apparent than in food production, which directly affects what we put into our bodies, and the broader consequences of that choice.

Many farmers selling pigs and chickens to giant vertically integrated suppliers no longer enjoy walking across green pastures on their farms with their children in search of butterflies, wildflowers, and frogs like I did with mine. The consolidation of meat buyers has left poultry and pig farmers with little choice but to accept contracts that turn them into low-cost input suppliers, not intelligent, creative, and adaptive stewards of the land. Their knowledge and culture have no value in cookie-cutter operations where buyers—who fully control vertically integrated production—dictate how their livestock should be raised. Practices that imprison livestock in crowded metal cages are designed to produce cheap protein ingredients for global trade, not to husband the land or sustain local communities.

The Green Revolution gave rise to more intensified animal production because the grain fed to them was made artificially cheap. When we tried to sell our organic cheese and meat at farmers' markets, our customers would often ask why our products were more expensive than the meat and cheese in the store. I'd explain that we didn't benefit from subsidized corn and soy to produce our grass-fed organic meat and milk, so they were paying the real cost of its production.

Conversely, when someone purchases a feedlot-sourced steak from a grocery store, the animal has been fattened on subsidized grains that the customer has helped pay for with their tax dollars. The customer had already paid for part of that steak. This certainly made selling organic grass-fed products much harder.

The young families who I most wanted to enjoy our healthy meat and cheese often couldn't afford the price we needed to charge just to cover our business expenses.

The stranglehold big meatpackers have on the USDA has also led to weakened meat-labeling laws. The USDA has pulled a curtain of confusion that obscures what goes on before food arrives on store shelves, making it harder for consumers to make informed choices. For example, the regeneratively raised ground beef from White Oak Pastures in Bluffton, Georgia must compete with cheaper imported beef that is not required to be labeled with its country of origin.

The steady supply of subsidized corn, wheat, and soy has enabled agribusiness to flood grocery store shelves with cheap, highly processed food loaded with carbohydrates, sugar, and unhealthy fats. A study published in The Lancet exposed the rise of a "double burden of malnutrition"—a simultaneous epidemic of hunger and obesity in low-income countries. Researchers attributed this to the restructuring of food systems in favor of global food marketing giants, displacing local markets that once provided diverse grains, fresh fruits, and vegetables.[358] Is this cheap food system really benefiting most people? What is its true cost?

## Consuming Our Grandchildren's Future

Mother Nature is a big loser in the race to overproduce crops. In recent years, soil erosion and the depletion of water sources have reduced humanity's ability to grow food.[359] American farm subsidies continue to encourage overproduction of corn and soy, crops responsible for most domestic soil erosion, weakening the country's wealth in natural capital.

How can consumers slay the Goliath that agriculture has become—a system built not to nourish, but to churn out cheap ingredients for junk food and ethanol? How do we reclaim control over domestic food production? Austin Frerick has extensively researched how agricultural policies have helped agribusiness limit our food choices since the 1996 "Wall Street Farm Bill." In his book *Barons: Money, Power, and the Corruption of America's Food Industry*, he offers commonsense steps to rewrite our food future. Restoring antitrust laws gutted by corporate lobbying is a good start. So is crafting policies that promote competition and offer healthier choices. If we care about the global impact of our diets, there's every reason for society to shift subsidies away from industrial giants and toward regenerative

**358** Barry M. Popkin et al., "Dynamics of the Double Burden of Malnutrition and the Changing Nutrition Reality," *The Lancet* 395, no. 10217 (2020): 69, https://doi.org/10.1016/S0140-6736(19)32497-3.

**359** David Pimentel and Michael Burgess, "Soil Erosion Threatens Food Production," *Agriculture* 3, no. 3 (2013): 443, https://doi.org/10.3390/agriculture3030443.

systems. Frerick calls for "putting animals back on the land" and phasing out polluting industrial operations, especially in his home state of Iowa where the very best American soils are degrading on a widespread scale.[360]

## Saving Black Gold for Our Grandchildren

Fossil fuels and the products made from them, such as chemical fertilizers, enabled the Green Revolution to push food production beyond the natural carrying capacity of the land and seas. It took 250,000 years for *Homo sapiens* to reach one billion people. In just the past 200 years, we added seven billion more people thanks to a one-time infusion of energy-dense fossil fuels.[361] In doing so, we decoupled our economies from the more enduring and dependable energy of the sun.

By wasting fossil fuels today in our excessively consumptive lifestyles, we limit their availability for the near-future generations who will desperately need it for essentials like food, shelter, and mobility. Today's generations would be considered wise ancestors if we conserved fossil fuels by using them judiciously to build quality, long-lasting infrastructure. Instead, we've been tempted to squander them in a frenzy of overconsumption of products that don't last.

By subsidizing fossil fuel extraction and ignoring the economic cost of pollution, policymakers have made these valuable resources artificially cheap, so consumers waste much more. Cheap fossil fuels have allowed the production of goods to swell global per capita consumption to 115 times as much as we consumed in the 1800s, enriching a wealthy few while fueling planetary overshoot.[362] Although fossil fuels appear cheap, their use has hidden costs that fall heavily on the communities where they are extracted. Wise stewardship of fossil fuels is humanity's best chance for a soft landing into the Restoracene.

## Inequality

Inequality is deeply entwined with planetary overshoot. The 38 countries in the Organization for Economic Cooperation and Development (OECD), including the US, Australia, Canada, Chile, Costa Rica, Japan, Mexico, and most European countries, are responsible for most of the global consumption that has catapulted humanity beyond ecological limits. In 2010, these countries accounted for

**360** Austin Frerick, Barons: *Money, Power, and the Corruption of America's Food Industry,* (Island Press, 2024) 178–79.

**361** Merz et al., "World Scientists' Warning: The Behavioural Crisis Driving Ecological Overshoot," 5–6.

**362** Will Steffen et al., "The Trajectory of the Anthropocene: The Great Acceleration," *The Anthropocene Review* 2, no. 1 (2015): 81–98, https://doi.org/10.1177/2053019614564785.

74 percent of global GDP, an indirect measure of consumption, despite comprising only 18 percent of the world's population. Considering the multiple crises all humanity now faces, how much we consume is a point of agency because consumption is something we can change today.

## Extinction Redemption in Perspective

The Quaternary Megafaunal extinction between 15,000 and 12,000 years ago caused such a shift in the Earth's ecosystems that megafauna biomass took more than 9,700 years to recover to pre-extinction levels.[363] The extent to which humans have appropriated the resources of wild megafauna for just us and our domestic animals is extreme. Once distributed across over 300 species, megafaunal biomass is now concentrated in humans and our livestock, making for a far less majestic and resilient ecology.

Wild megafaunal numbers have continued to decrease faster. Based on genomic modeling, wild megafaunal numbers have decreased by 89.4 percent in the last 50,000 years as humans spread out of Africa. Some scientists think at least 660 million individuals were lost, or 8 million tonnes of biomass.[364] Because humans continued to take over the habitats of the wild species they exterminated during the Late Pleistocene Extinction, no wild replacements have started to evolve, and they will not be able to if humans keep appropriating all the ecological niches for ourselves.[365]

Yinon M. Bar-On, Rob Philips, and Ron Milo studied the historical abundance of wild animals and humans. They estimate the biomass of all wild terrestrial mammals before the Quaternary Extinction Event was 20 million tonnes of carbon biomass (**Figure 15.1**). Marine mammal biomass was fivefold today's biomass at another 20 million tonnes of carbon. Humans and our livestock biomass now total 170 million tonnes.[366] This means we have little choice but to use livestock and ourselves to restart land productivity where we have degraded it. We must try to return the miracle of photosynthesis to every place it once prevailed with the help of the only megafauna now abundantly available.

---

**363** Anthony D. Barnosky, "Megafauna Biomass Tradeoff as a Driver of Quaternary and Future Extinctions," *Proceedings of the National Academy of Sciences of the United States of America* 105, (2008): 11545, https://doi.org/10.1073/pnas.0801918105.

**364** Juraj Bergman et al., "Worldwide Late Pleistocene and Early Holocene Population Declines in Extant Megafauna Are Associated with Homo sapiens Expansion Rather than Climate Change," *Nature Communications* 14, no. 1 (2023): 7679, https://doi.org/10.1038/s41467-023-43426-5.

**365** Barnosky, "Megafauna Biomass Tradeoff," 11545.

**366** Yinon M. Bar-On et al., "The Biomass Distribution on Earth," *Proceedings of the National Academy of Sciences of the United States of America* 115, no. 25 (2018): S1, 89, fig S5, https://doi.org/10.1073/pnas.1711842115.

Since the Industrial Revolution, the infusion of fertilizer and mechanical work provided by fossil fuels into the global energy budget has grown more food and allowed humans and our livestock to grow to levels six times higher than the biomass of megafauna the Earth could support naturally.[367]

Human population levels higher than preindustrial levels will be impossible to support when fossil fuels become exhausted. This reckoning could come soon.[368] Humanity needs to manage our own population to reflect the natural productivity of the Earth's agroecosystems by relieving poverty, educating everyone, and using humane forms of birth control.

I am "pro-life." For me, all life created by God is sacred, be it human or non-human. I also think abortion extinguishes the most vulnerable life on the planet. If modern birth control technology were readily available to anyone who wanted it, many abortions could be prevented. Abortion is not something women would want to endure if they could avoid it. If propaganda promoting endless growth on a finite planet continues however, sensible and humane pathways to rebalance human population will not be possible. The immense suffering endured by so many beings on our planet will continue. I have alluded to signs of resource grabbing by the richer societies among us in this book. This can only accelerate if we let our numbers explode beyond the carrying capacities of our homelands.

The time is overdue for humans to embrace population deceleration. Education in ecology and family planning can humanely slow population growth, lowering the demand for livestock and companion animals, too. Today, the populations of just our dogs exceed the biomass of all wildlife.

People are out of harmony with our home planet. Population deceleration could help us achieve better quality of life for our future families, who will need a healthy, productive planet. It will be far wiser to plan a soft landing into the Restoracene, instead of a crash landing here in the Anthropocene.

---

**367** Barnosky, "Megafauna Biomass Tradeoff," 11546.

**368** Barnosky, "Megafauna Biomass Tradeoff," 11546.

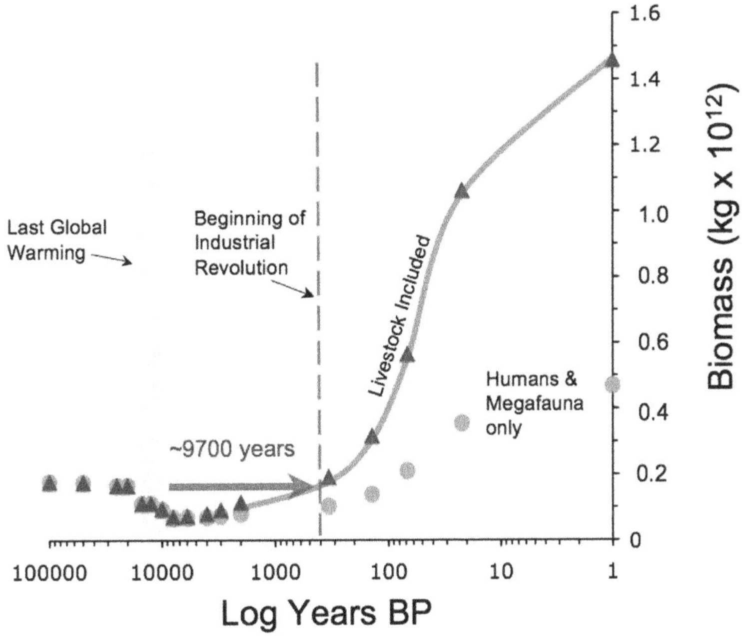


**Figure 15.1. The biomass of megafauna over the last 100,000 years.** Vertical bar indicates the timing of the YD-Holocene climatic event that led into the current interglacial.

Reproduced from Barnosky, et al., 2008.

## People Power

There is no Planet B. The time has come to stop allowing marketers to convince us to ignore looking after our home planet with fantasies that space exploration will come up with a new home for humans. This is convenient and tempting but only perpetuates a failed colonial culture of discovery for exploitation that has nearly exhausted all lands habitable for humans on our beautiful planet. Compared to planet Earth, Mars sucks! You don't have to be complicit in the wasteful consumerism that is destroying your own planet.

Our species will learn far more about rebuilding fading ecosystems when we regain our position as an integral part of nature in the Restoracene. As people become highly educated in ecology and more poor people find well-paying job opportunities restoring green landscapes that can feed them reliably, all families will finally have better choices around family planning. In the meantime, the horse has left the barn as far as a sustainable population is concerned. The human population will grow to 10 billion before it starts to decline, so our only opportunity to address the many areas of planetary overshoot is to stop buying and throwing away so much unnecessary stuff!

We can start now to reduce waste by treating fossil fuels as a precious resource to be conserved wisely. We can start using them to restore deserts and regreen our planet to keep it productive and cool instead of producing mountains of unnecessary stuff out of plastic and moving it around the planet using even more fossil fuel. Now is the time to use the energy-dense fuels we have left to build durable and efficient mass transit infrastructure so people will not forever be locked into buying their own car to get to work. Car ownership traps people with the burden of paying for their associated suite of constant consumption of gasoline, diesel, tires, oil, lubricants, batteries, replacement parts, antifreeze, car washes, roading, taxes, permits, et cetera. If politicians worked in the interest of today's and tomorrow's people instead of the corporations that sell all these products, we would have public transportation infrastructure that was fit for purpose already.

## Chapter 16

# The People of the Restoracene

Two million years of tight-knit cooperation within our tribes, like the single-mindedness of a world-class soccer team, allowed humans to thrive in the game of evolution. But in just the past 200 years, the capitalistic global trade model has divided communities and extended families. Our human connections used to provide the support and comfort individuals need. Like most mammals, we don't feel safe or happy alone. Yet global markets benefit from breaking communities up into smaller units—individuals or nuclear families—so they can sell more TVs, cars, kitchen appliances, and so on.

By participating in excessive consumerism, we permit corporations to bulldoze values, social safety nets, and community-strengthening customs that a kaleidoscope of cultures developed over thousands of years. We have replaced these nuanced and useful customs with a homogenized but glitzy orgy of branded consumption. To acquire all the products that corporations convince us we need, we work more and spend less time with our loved ones. Grandparents, who were once integral to the household by sharing wisdom, acceptable social behavior, and cultural richness with their grandchildren, are now distant. Instead, their grandchildren are soothed by electronic screens that are not designed to develop kind, capable members of the community, but misinformed, gullible consumers.

Meanwhile elders, who once were safely enveloped in extended families that were knit into a supportive community of longtime neighbors, now find themselves in smelly retirement homes waiting in vain for a visit from a familiar face. What price are families in the Western world paying for the privilege of enriching shareholders who don't care about us? Where should our loyalty lie? Who is going to stand by us when times get tough or (human) predators circle? If our economies need reimagining, so too does the fabric of Western society. We must rebuild strong, supportive communities that value human and ecological collaboration for long-term survival. Resilient communities don't just happen.

Reconnecting takes a bit of work. I learned this when fracking came to town, and our community decided we would not be shoved into the polluted, impoverished future fossil fuel companies were creating in other US states. We decided to take ownership of our own story.

## Water: The Lifeblood of Raindance Farm

My cousin Summer liked visiting Raindance Farm for a couple of weeks during her school vacations when she was a teenager. I recall one hot day when we went tubing together: climbing into big old tractor-tire tubes to float down the Cherry Valley Creek from one end of the farm to the other. The trip took two hours because the creek leisurely wound around many horseshoe bends on our 135-hectare farm. We enjoyed floating past drooping willow trees and sturdy beaver dams, as crickets and birds chirped in the adjacent grasslands and forests.

Soon after we had settled into the relaxing pace of the lazy summertime river, a noisy duck startled us as it fled over the water. Her wake on the glassy water drew our eyes to a young coyote standing on the bank. He was looking just as surprised to see us as we were to see him. As if he was locked in a photographer's frame, he stood motionless, holding the duck's freshly caught partner in his mouth. That moment captured the wildness that still pulsed in the riparian heart of our farm.

My daughter Keira and her friends would always stop at some point while tubing down the creek to paint their bodies with dark, silty creek mud. The mud had a magic quality that caused uncontrollable laughter. Another favorite farm activity after a hot, dusty day making hay was jumping off the old wooden-decked bridge into the creek.

When our children were younger, we'd gather with neighbors on the streambank, sipping cool drinks while the children swam, fished, and searched for crayfish under stones. Our nearest and dearest neighbors had mowed a meandering path through their three-hectare hay field to the creek behind their house. At evening get-togethers, Chris lit the way with tiki lamps so we could enjoy evenings sitting creekside while the children played. The evenings were mosquito-free thanks to many aerobatic bats.

One Mother's Day, my two teenagers paddled me down the creek to meet the nearby Susquehanna River headwaters in our wood strip canoe. I sat in the middle like a river queen, stopping along the way to pick the wild leeks we called ramps. We loved the creek and all the nature around it! So, when the health of the creek was put at risk, I couldn't stand by. The Cherry Valley Creek was more than a waterway; it was the elixir of life, connecting people and wildlife on Raindance Farm.

## Ripping Back the Curtain on Corporate Spin

In 2010, the fracking industry set its sights on Otsego County. Fracking (or hydraulic fracturing) is a process where rock deep underground is cracked open to release methane or natural gas. One of my neighbors, Kim Jastremski, told me that a farm property three miles upstream from us had sold the rights to develop their land for fracked gas. A nearby farm at the headwaters of the Susquehanna River in Cooperstown had done the same.

The story fossil fuel companies were trying to sell to our community was that we would all get rich from gas "development." But we had been forewarned. People living in fracking zones in Pennsylvania had already alerted us to a different reality being created by gas companies—broken promises, underpaid landowners, and polluted aquifers.

I had been informed of other associated dangers by one of my organic dairy farming friends, who had developed a beautiful family farm in Pennsylvania. His son's family had joined him in running the place, fulfilling their American dream. Then a gas company built a pipeline pumping station across the road from their farm as part of a scheme to export their valuable local natural resources on the cheap to overseas ports, and shattered their dreams. Diesel generators ran day and night, polluting the air with fumes, light, and noise. His son's family was forced to move away for health reasons. The gas companies had stolen their right to enjoy their property in peace and pursue happiness.

My neighboring farmers probably gave away the rights to the fossil fuel riches under their farms for the same deal I was offered: 12 percent of profits after all expenses were paid. This was a far worse deal than the 50/50 deal my local logger offered when he selectively harvested fully mature trees on our farm before they declined from old age. In the meantime, the gas company wanted to lease the whole farm for just six dollars an acre per year. The lease gave the fossil fuel company audacious privileges to cut down forests and build roads, worker housing, and industrial infrastructure, including ponds for storing toxic fracking fluid. I refused to sign this lease and told the gas company representatives not to come back. After all, some residents in Pennsylvania could literally ignite the stuff coming out of their kitchen faucets because their water aquifers were contaminated by fracking activities.[369]

**369** Scott Cannon, "Water on Fire from Fracking Gas Drilling," posted by Gas Drilling Awareness Coalition, accessed January 15, 2026, https://youtu.be/4LBjSXWQRV8?si=CCxSNN4qJnNNu7XG.

We had delicious water, and we wanted to keep it that way. It takes hundreds of truckloads of chemical and diesel-infused water to crack open and keep open the shale rock beneath each natural gas well. Fossil fuel companies had lobbied for exemptions from disclosure laws so we could not find out what chemicals were in the fluids. The chemicals behind this curtain of confusion couldn't have been harmless, or why would the corporations hide the information? The Pennsylvania Department of Conservation has documented 444 water wells contaminated by oil and gas activities as of January 2025.[370]

In response to the threat to our livelihoods, Kim, myself, and five other concerned parents in the town of Middlefield formed a grassroots organization to protect our local waters. We called ourselves "Middlefield Neighbors." We had no intention of allowing an outside industry to destroy our natural capital or our quality of life—especially the clean, fresh water we enjoyed in abundance. In no time our numbers grew.

## Our Fight for Self-Determination

Cooperstown is at the headwaters of the vast Susquehanna watershed. Pollution flows downhill, so we have some of the cleanest water in the Northeast. We wanted to protect it not just for ourselves, but for everyone downstream who drew on it for drinking water.

As a farmer, I was alarmed by a story from Pennsylvania where cattle had died after drinking contaminated water that had leaked from a containment pond on a fracking pad into their pasture.[371] Where would all the toxic wastewater from fracking go? The plastic-lined ponds couldn't hold it forever and were bound to overflow, especially with the more frequent flooding we were experiencing. As an organic livestock producer, the threat of contamination of the major tributary flowing through my farm was an existential economic crisis. When I grazed pasture on the other side of the Cherry Valley Creek from our barn, I used a solar-powered pump to fill a water trough with river water for the cows to prevent them from entering the creek itself for a drink. The water needed to be pure. Also, when the creek flooded, water would overflow into my fields. If the containment ponds upstream were also overflowing because it was raining hard, my pastures would be contaminated. My customers would not want toxic chemicals to work their way

370 "Water Supply Determination Letters Letters," Pennsylvania Department of Environmental Protection, accessed January 15, 2025, https://files.dep.state.pa.us/OilGas/BOGM/BOGMPortalFiles/OilGasReports/Determination_Letters/Regional_Determination_Letters.pdf

371 Elizabeth Royte, "Livestock Falling Ill in Fracking Regions," NBC News, Nov. 30, 2012, https://www.nbcnews.com/news/ investigations/livestock-falling-ill-fracking-regions-flna1c7336300.

up the food chain and into our milk or Sun Cheese. The Park Slope Food Co-op in Brooklyn had already declared they would not be buying produce from farms in fracking areas. Fracking would put my farm out of business.

## Wresting Back Autonomy Over Our Future

Over 80 percent of residents in our peaceful, safe, rural town told us in our community phone survey that they didn't want fracking to come to town. Citizens told us they valued our bucolic landscape and wanted the same for their children. Compressor stations, heavy truck traffic, and gas wells flaring day and night would not support the peaceful lifestyle we all valued.

Middlefield Neighbors took further action. After dinner, when people were relaxing in front of their TVs, we knocked on doors and asked if they would sign a petition to the town board to zone out heavy industry, including fracking. Hundreds signed. Many of these residents were lovers of the outdoors and enjoyed fishing, hunting, hiking, and skiing. More than three quarters of those who responded stated that they did not want fracking in the areas of their township that were zoned for agriculture.

Humanity's best defense against the corporate spin that pushes for overconsumption is holistic context creation. Knowing what kind of future you want and protecting it is an important strategy. Our family's holistic context included a landscape rich in clean water and sufficient natural resources where our descendants could eat well and live comfortable, peaceful lives. It went against our holistic context to allow oil and gas corporations to come into our township and permanently pollute our aquifers to rapidly extract our resources for export. There was enough natural gas in the shale under our town to last our local community hundreds or thousands of years. Yet a Canadian company planned to suck it all out in just five to 10 years, with none being used directly by any of us or our descendants.

Most of the profit from gas development would go to faraway shareholders. Even the jobs created to do the work would be filled by out-of-town industry specialists. The fossil fuel giants were writing a script we wanted no part of. Our "guiding star" vision of the future we wanted was our secret weapon. We wished our grandchildren to have the same opportunity to go swimming, fishing, and canoeing on the river and to be able to make a good living from agriculture. But we also wanted them to have access to the highly concentrated energy in natural gas that they would need to build essential infrastructure and keep warm in our cold Restoracene winters 100 and even 1,000 years from now.

## Resisting Lobbying

If you look at the websites of lobbying companies, they boast about their "relationships" with lawmakers and the special access they have. The voices of ordinary people trying to stick up for nature struggle to be heard. Citizens have lost much autonomy. I discovered this firsthand when I traveled to Albany, the New York State capital, in 2011 with a delegation from Middlefield Neighbors to share our concerns about fracking with Governor Andrew Cuomo.

Our group included a cross-section of the community, including people knowledgeable about the proposed fracking. We had all shared with each other what we had learned about fracking, and some of us had visited communities in Pennsylvania to get firsthand accounts of how it was affecting people there. Ron Bishop, a biochemistry professor at the State University of New York at Oneonta, helped educate us about the dangers of commonly used fracking chemicals. We were joined by Louis Allstadt, a former Mobil vice president who was alarmed by weak regulations and the possibility of methane leaking from fracked wells for hundreds of years into the future. Kelly Branigan, a nurse who had researched the health effects of fracking in other regions, also came with us, as well as homemakers who were worried about their families' health. I was asked to speak from a farming and economic perspective.

Although our group was a snapshot of the voters Governor Cuomo was supposed to represent, we were told the governor was busy and we would only be seen by his Deputy of Energy, Thomas Congdon. Upon entering the State Capitol parking garage, however, we couldn't help but notice the rows of white vehicles emblazoned with fracking company logos. It was clear: The game was rigged. Money was buying access and the people's voices were not going to be heard by the governor. Our group of concerned citizens, who had quickly become friends and helped inform each other, were more determined than ever to not allow fossil fuel corporations to change our lives for the worse.

Meanwhile, a like-minded group we coordinated closely with on the other side of Cooperstown's Otsego Lake, led by Julie Huntsman, was carrying out a parallel campaign. On June 14, 2011, Middlefield became the second town, after our neighbors in Otsego, to zone out heavy industry like fracking at the local level, asserting our citizens' right of self-determination. Our local town board's decision was based on New York's "home rule" provision that guarantees town laws trump those of the state. Eventually, 55 towns in New York followed suit. Fossil fuel interests fought us in court but in June 2014, the New York Court of Appeals upheld our town's right to regulate fracking through home rule.[372] The political winds had changed. Governor Cuomo, to his credit, subsequently banned fracking in New York State in December 2014.[373]

Although many people want to see an end to fossil fuel use altogether, there is no sign of that happening. Fossil fuel use has only steadily increased since the public became aware of climate disruption decades ago. I hope the natural gas under Middlefield Township remains underground to benefit future generations in the Restoracene, when it will be far more valuable so it will be used conservatively. To send it away now as a cheap export to other countries only to drive wasteful consumption and enrich a few already wealthy people would steer us away from the stable future I want. My lesson from the fracking battle gave me hope that resistance, initiated locally but applied widely, can change government policy. Likewise, local changes to make land greener, applied widely on landscape scales, can cool regions, if not the whole planet, as we evolve into the Restoracene.

## The Long Tentacles of Fossil Fuel Pollution

For me, the fracking issue was never just about water. Trouble was also brewing in the air. As the fracking boom took off a couple of hours west of my farm in Pennsylvania, I started experiencing intermittent chest pain and coughing. I started paying closer attention to the quality of the air around me, and I could see that on the days when I had symptoms, the air was visibly smoggy. I soon knew before I even opened my curtains in the morning if the hills that our valley nestled amongst would appear clearly or be veiled in haze. If I had chest pain and I was coughing, the hills would be shrouded in haze: a combination of pollution and moisture. I had become the canary in the coal mine.

When I checked the Air Quality Index online, I was shocked to see that the air in my nearest town, Cooperstown, often had higher levels of polluting particulates than the air in the Bronx, four hours away. We were far from big cities or dirty industry, so how could our air be worse than in the city? Could there be a correlation between this bad air and the fracking boom in Pennsylvania? The answer was in the wind. The prevailing southwesterlies coming from the direction of fracked Pennsylvania farmland could be carrying airborne pollutants our way.

**372** Daniel Wiessner, "N.Y. Top Court Says Towns Can Ban Fracking," Reuters, July 1, 2014 https://www.reuters.com/article/ markets/n-y-top-court-says-towns-can-ban-fracking-idUSL2N0P61BT/.

**373** Thomas Kaplan, "Citing Health Risks, Cuomo Bans Fracking in New York State," *New York Times*, Dec. 17, 2014, https://www.nytimes.com/2014/12/18/nyregion/cuomo-to-ban-fracking-in-new-york-state-citing-health-risks.html.

The air quality in many rural towns in the Western US has deteriorated because of natural gas and oil extraction. A study from the Uinta Basin in Utah, formerly one of the most pristine environments in the country, revealed that the total emissions of volatile organic carbon compounds were equivalent to the annual emissions of about 100 million cars! It also found a strong causal link between oil and gas emissions, buildup of air toxins, and significant ozone formation near the atmospheric surface layer.[374]

## The Metastasis of Fossil Fuel Pollution

Although fracking was banned in New York for the time being, we couldn't escape one of its most pervasive externalities: air pollution. From 2014 until I left the farm in 2016, my respiratory symptoms had worsened to the point that some days I had chest pain, lacked energy, and had to spend most of the day on the couch. Over the previous two years, my doctor had prescribed several asthma medications. Nothing worked. He ruled out asthma but never quite put his finger on the cause of my symptoms before he recommended I quit farming. Although it broke my heart to sell what felt like part of me, I couldn't bear the thought that one day I would wake up too sick to look after my cows who had sustained my family for so many years.

I suspect the cause of my symptoms was tied to the dirty air blowing in from the fracking boom taking place to the southwest of us, because I felt ill when the air looked visibly dirty. The closest Pennsylvanian counties endure a disproportionate burden of fracking in the state. Despite all my efforts to resist the fossil fuel industry, they had rewritten my story.

I only recently looked closer and discovered that health data supports my suspicions. Chronic lower respiratory disease (CLRD) is defined by the US Centers for Disease Control and Prevention and the World Health Organization as encompassing four major diseases—chronic obstructive pulmonary disease, chronic bronchitis, emphysema, and asthma—and in 2019, it was the fourth leading cause of death in the US.[375] The hospitalization statistics from the

**374** D. Helmig et al., "Highly Elevated Atmospheric Levels of Volatile Organic Compounds in the Uintah Basin, Utah," *Environmental Science & Technology* 48, no. 9 (2014): 4707–15, https://doi.org/10.1021/es405046r.

**375** YC Lee, KY Chang, S Sethi, "Association of Chronic Lower Respiratory Disease With County Health Disparities in New York State," *JAMA, Network Open* 4(11): e2134268, (2021): intro, doi:10.1001/jamanetworkopen.2021.34268.

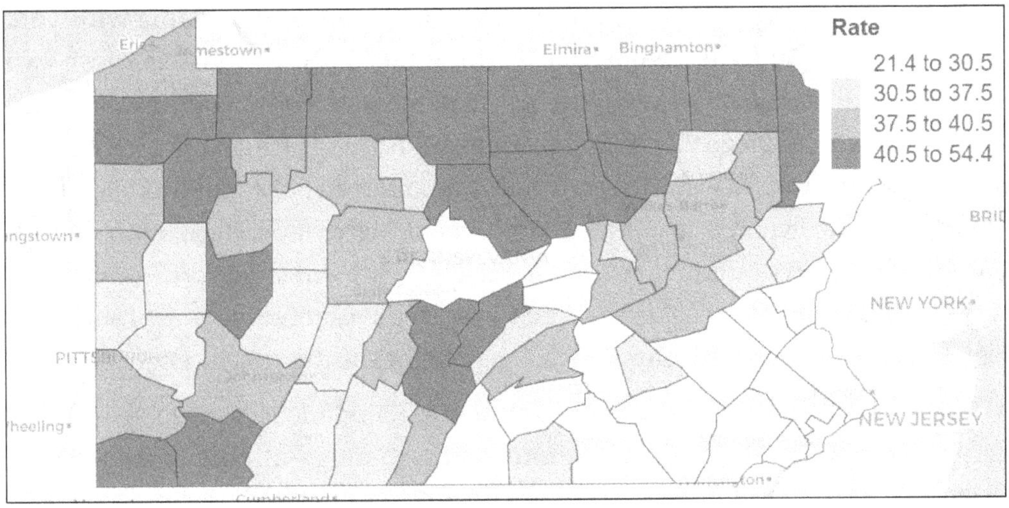


**Figure 16.1 Chronic lower respiratory diseases age-adjusted death rate per 100,000 people, 2019–2023.**

Pennsylvania Department of Health website.

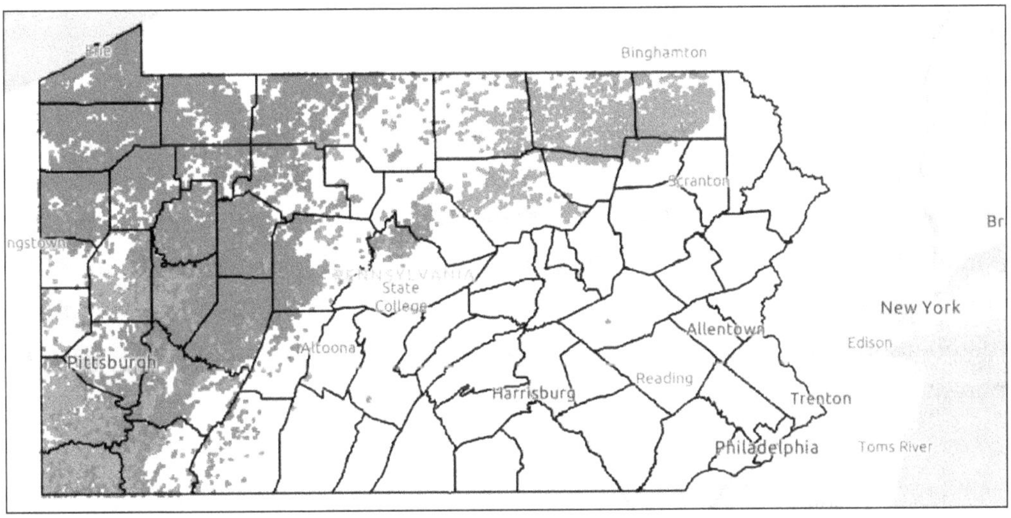


**Figure 16.2 The intensity of natural gas production in Pennsylvania in 2025.**

FracTracker Alliance website.

Pennsylvania County Health Profiles published online and updated in February 2025 show that the Northern Tier of Pennsylvania and most southwestern counties experience the highest rates of chronic lower respiratory disease (**Figure 16.1**).[376] Unsurprisingly, this mirrors the intensity of oil and gas production (**Figure 16.2**).

376 Pennsylvania Department of Health, "County Profiles, Hospitalizations, Maps, Chronic Lower Respiratory Disease," updated February 2025, accessed August 7, 2025, https://www.pa.gov/content/dam/copapwp-pagov/en/health/documents/top-ics/healthstatistics/vitalstatistics/countyhealthprofiles/documents/current/maps-hospitalizations.html#chronic-lower-re-spiratory-diseases-1.

Then, I compared Bradford County in northeast Pennsylvania with York County in the southeast, where no fracking takes place, because these two counties have identical rates of smoking (15 percent of the population). Bradford County, which produced 20 billion $m^3$ of gas in 2015, had an age-adjusted chronic lower respiratory disease (CLRD) hospitalization rate in 2023 that was 58 percent above the state average, and had a 119 percent higher hospitalization rate for chronic obstructive pulmonary disease (COPD)![377] Meanwhile, York County with little impact from fossil fuel extraction but with the exact same rate of smoking, experienced rates of CLRD that were 24 percent below the state average.[378]

The three Pennsylvanian counties closest to Raindance Farm where fracking is taking place—Susquehanna, Bradford, and Wyoming—all share elevated lung disease rates. Millions of Americans are nonconsenting test subjects in a grand experiment. And these are just the early signs. What long-term health costs will be borne by people raised in the path of fossil fuel pollution in the coming decades?

## Building Resilient Communities

Despite undercurrents of "not in my backyard" attitudes, Middlefield Neighbors became acutely aware of the bigger picture. We knew we had to reduce our fossil fuel use to prevent worsening climate heating and to make our community and our country more resilient. If we continued the American status quo of wasteful consumption of energy, we would be complicit in the destruction of water and ecosystems of other communities. Many members of Middlefield Neighbors installed solar panels on their homes. At Raindance Farm we added a 13-kilowatt photovoltaic array on the barn roof, installed by a local Middlefield company.

---

**377** Pennsylvania Department of Health, "County Profiles: Bradford: Hospitalizations," updated February 2025, accessed August 7, 2025, https://www.pa.gov/content/dam/copapwp-pagov/en/health/documents/topics/healthstatistics/vitalsta-tistics/countyhealthprofiles/documents/current/bradford.html.

**378** Pennsylvania Department of Health, "County Profiles: York: Hospitalizations," updated February 2025, accessed August 7, 2025, https://www.pa.gov/content/dam/copapwp-pagov/en/health/documents/topics/healthstatistics/vitalstatistics/countyhealthprofiles/documents/current/york.html.

## Meaningful Work

By using our holistic context to guide important decisions like investment in renewable energy, we supported the development of local jobs and new businesses. Unlike the fleeting service-sector employment that comes with a fracking boom, jobs in renewable energy had staying power. As a rule, drilling companies hire out-of-state, industry-specific workers. Our solar installer was local, committed to the community, and a reliable, good neighbor. That's what strengthens communities in the long term.

There were other signs that the tattered threads of community were being mended. The coming together of our townspeople renewed popularity in the local farmers' market as our neighbors rallied around the local food movement. Young entrepreneurs started up a popular farm-to-table café called Origins Cafe. They sourced ingredients from local farmers, including grass-fed organic sausage from Raindance Farm. It became a hub for live music, local events, and good conversation. Reconnecting with each other is essential if we hope to reconnect with the natural world we're meant to protect.

## Chapter 17

# Welcome to the Restoracene

In the Restoracene, people will see themselves differently. Presently, corporate interests want to define us as two-faceted beings, workers on one side of the coin and consumers on the other, whose purpose is to enrich them both coming and going. Those who manage to amass more money than others so they can consume more, are put on social pedestals even though they are the worst ancestors. Now, many people turn inward, placating shallow, insatiable egos that we cultivate on our phones. We spin our wheels faster to acquire more stuff that never seems to satisfy us. It never will because it is not what we need. As we evolve into the Restoracene and begin to revere abundant life and connect meaningfully with our community, we will have everything we need. We will become whole again.

Once, we lived in harmony with our native ecosystems, as symbionts. We helped each other thrive while we explored and learned together. The most generous amongst us were revered, respected, and loved. Giving felt better than taking because it strengthened our life-sustaining social network, just like plants increase their chances of success by sharing the riches they gather from the golden sun with other species. We were once embedded in the self-renewing circle of life that cradled our highly successful tribes. We were held spellbound every day by nature's complexity, beauty, danger, and adaptability. We lived from the interest of the natural capital surrounding us, and nature efficiently recycled all our waste. We looked up and outward in awe at how very firmly we were connected to exceptional abundance and each other.

We can be these people again in the Restoracene if we can learn from the past. The Māori have a whakataukī, similar to a proverb but with much deeper meaning. *Kia whakatōmuri te haere whakamua:* "I walk backwards into the future with my eyes fixed on my past."[379]

**379** Lesley Rameka, "Kia whakatōmuri te haere whakamua: 'I Walk Backwards into the Future with My Eyes Fixed on My Past,'" *Contemporary Issues in Early Childhood*, 17, no. 4 (2016): 387, https://doi.org/10.1177/1463949116677923.

## Paradigm Change

What persuasive force could possibly push *Homo sapiens* to evolve quickly to make wise decisions amid the noise and complexity of modern life? We've known about our role in climate disruption for at least 45 years, but fossil fuel use keeps rising. Environmental outcomes continue to get worse. What might spark a genuine shift?

Could it be a change in how we see ourselves—as conscious, intentional ecosystem engineers? If we own the likelihood that our species has been influencing the climate for at least 64,000 years, can we then also accept that we have the agency to influence it for the better through how we manage land, seas, business, economies, and our own lives? Will we be inspired to act by the hope this paradigm shift reveals?

To reverse biodiversity loss and declining planetary integrity, we need a clear picture of where we are heading. We need to see in our mind's eye what a healthier future can look like. The teeming landscapes our ancestors enjoyed, before they limited ecosystem productivity, can be our guide. Although we may never recover the full species richness our ancestors stole from us as they arrived in new continents, we can work with what we have left. We can restore abundance and activate evolution on our planet once again. Evolution is very slow, so we'd better get started. Luckily, nature has provided templates we can learn from to replenish natural capital in the form of soils, fish stocks, fresh water, and more. The farther back we look, the farther into the future we can see.

## Restoring Life with Purpose

Thanks to holistic management practice, degraded ranchlands, deserts, and even former mining sites are being transformed into green, life-giving biomes. Ranchers on different continents, like Roland Kroon in South Africa, Chris Henggeler in Australia, and Alejandro Carrillo in Mexico, are proving how holistically managed livestock and wildlife can be a tool to regrow green cover, even in deserts. This vegetation cools the land and restores water cycles quickly and cheaply compared to any technology.

The lessons holistic management practitioners like us are learning will be valuable to the billions of people who make up the greatest but most wasted natural resource on the planet: humans who do not have meaningful, well-paid livelihoods. Large tracts of land on our planet are degraded and missing the megafauna they need to thrive—and that includes people. Climate scientist Walter Jehne estimates that two billion out of the Earth's five billion hectares of desert worldwide are restorable—enough to generate a massive cooling effect on the climate. These lands are waiting for people to return and regreen them. With so much restoration that needs to be done, can the billions of under-resourced people on the planet be empowered to become the heroes of the Restoracene?

Millions of meaningful jobs can be created for the underemployed people who are fleeing degraded homelands in search of opportunity. The work that needs to be done includes land and ocean restoration work itself, the teaching of ecology and resource management required, the development of technology to assist restoration, and the associated monitoring of biodiversity to vet results. Everyone on planet Earth would be better off if potential migrants had educational and job opportunities right at home to restore nature's green cloak. But where will the money come from to reward people willing to help make the planet more livable for everyone?

The money is there to do the good work required to begin the Restoracene. We just need collective will. This can manifest itself if we look forward to a better life of abundance instead of fighting over the scraps left after squandering so many of nature's gifts. According to the International Monetary Fund report, *IMF Fossil Fuel Subsidies Data: 2023 Update*, direct subsidies for fossil fuels to make them artificially cheaper to world citizens in 2022 totaled US$1.5 trillion. This is a vast sum, equal to the total GDP of Russia. But if the external costs placed upon humanity because of the use of dirty fossil fuels are also accounted for, such as health impacts from air pollution and damage from climate change, the total fossil fuel subsidy cost is $7 trillion, or 7.1 percent of global GDP.[380] Add to this the $2.7 trillion military expenditure in 2024 (2.5 percent of global GDP) according to the Stockholm International Peace Research Institute,[381] and it becomes clear that almost 10 percent of everyone's effort on the planet could be redirected to making the world a better place instead of polluting it and killing each other. We surely can imagine more humane economies using the tool of human creativity.

Global policy changes could redirect these $10 trillion and other appropriate money streams, such as fines for corporate polluting, and luxury consumption taxes, into a fund to reward the poorest two billion people on the planet for greening land and lifting biodiversity. Five thousand dollars or more could be available to each restorer annually. Dreams then become possible. For instance, poor villages of 100 people on degraded land could borrow $500,000 from the fund

**380** Simon Black et al., "IMF Fossil Fuel Subsidies Data: 2023 Update," IMF Working Papers 2023, 169, (2023):1, https://doi.org/10.5089/9798400249006.001.

**381** Xiao Liang et al., *Trends in World Military Expenditure, 2024* (Stockholm International Peace Research Institute, 2025), 1, https://doi.org/10.55163/AVEC8366.

to plant a food forest to feed themselves and produce agricultural goods to sell. As satellite technology verifies that trees and plants are increasing green cover every year over the ensuing 30 years, the people will be rewarded by the fund and can pay themselves for their effort to maintain and protect their newfound abundance as they pay back their loan from the fund. They will be incentivized to perform the work efficiently, so they have money left over from the rewards for other actions that allow them to move closer to their holistic context. For instance, they could build themselves a solar-powered water pumping station to irrigate and green even more land over the next decade, potentially earning more rewards for the next generation.

Schemes like this would jump-start local economies and build pride and ownership of place as they bring the Goliath of inequity to its knees. Economic activity will be defined by meaningful work instead of the manufacture and transportation of useless plastic goods around the planet. Best of all, people will be better able to feed themselves. Food and water security will be the incentive for people to take care of their projects and their land to sustain themselves going forward.

The knock-on effect of empowering the least resourced people on the planet will be massive for conservation. People that are not desperate for food and fuel will not be forced into the dangers of poaching or need to scour the forests for bushmeat and wood for charcoal to cook it. For these and thousands of other reasons, the widespread ecosystem restoration required for a livable planet is not possible without the restoration of equity and justice.

As neighboring villages also put local wisdom and holistic training to work to green their surrounding land, regions will see more rain and become even more productive. Aquifers will once again fill for the benefit of today's and future generations. Humanity's impact on landscapes will be reversed as people are empowered to grow food and fiber for themselves. But our economies will need to be reimagined.

## The Rise of a Solar Economy

Petrostates like Saudi Arabia, China, and the US that have vast tracts of desertified land in need of restoring have much to gain if citizens are paid to restore their countryside so future generations can be fed and watered in a post-fossil fuel future. Countries can leverage the last economically accessible supplies of fossil fuels to restore the functioning ecosystems that are required to replenish their depleted aquifers, build arable soils, and create cooler, livable landscapes.

Cheap fossil fuels to run irrigation systems and build the smart, fit-for-purpose infrastructure required will not be around much longer at the rate we are squandering them. Regreened productive lands will reduce the need for irrigation and will increase the true wealth of nations.

## Rewriting Your Future

I offered my regenerative farming story not as a how-to, but to inspire everyone to evolve to a holistic manner of reimagining and achieving their future. There are many ways to manage resources to support biodiversity and sequester carbon. Because each place is unique and has issues around restoration that are constantly changing, the people living in those places will know how to best restore their landscapes considering their unique social needs. Their elders can help. But all communities will need education in ecology, holistic management, and resources to get to the Restoracene.

## Finding the Grace of Pure Nature

You and I are connected to everything. John Muir wrote in his first book, *My First Summer in the Sierra,* in 1911, "When we try to pick out anything by itself, we find it hitched to everything else in the Universe." Who would have thought the return of blue butterflies, bluebirds, and bobolinks would have inspired me to understand my very purpose in life? I was born to be a symbiont in the self-organizing, mighty web of life. That insight will keep me busy with meaningful work for the rest of my life. Each time I teach a child to plant a tree, skip an unnecessary car trip, resist the e-commerce "buy" button, or save a life, no matter how small, I'm fulfilling that purpose. I am being the good ancestor I want to be. That gives me both hope and self-empowerment, two things becoming scarcer day by day.

What will life be like for our grandchildren in the Restoracene? It may be closer than we think. Nature has an extraordinary capacity to heal if enough critical mass still exists. And I believe it does—just. So, I wrote a letter to my great-grandchildren. I needed to think about what kind of ancestor I've been. They'll be here soon enough.

And they deserve the truth.

**Let the Restoracene Begin.**

**www.restoracene.com**

Dear Great-Grandchildren,

You're not born yet as I write this, but I am thinking about you.
I've tried to be a good ancestor; to leave the world better than I found it, so you could hear birdsongs, swim with dolphins, and feel soft spring rain on your face. I tried to grow my own food with mixed success and built deeper soil on my organic dairy farm.

But I was also greedy about my fossil fuel use, sometimes flying for frivolous reasons like to go surfing in faraway places. I'm sorry for using up fossil fuels that I know you need to stay warm in winter, make fertilizers to grow enough food, and build renewable energy infrastructure. Fuel and many goods are so expensive for you now because my generation wasted too much.

I hope the tigers my daughter loved, and all other species, made it to your generation. And the penguins too. The yellow-eyed penguins in New Zealand were nearly gone when I wrote this. My friends and I tried to help them. They didn't have enough fish to eat, so I stopped eating fish even though it tastes so good. Hopefully unsustainable fishing was stopped before it was too late so you can go fishing and catch something for dinner.

That was one of the things I wrote about in my book to try to help people understand that for tens of thousands of years our ancestors took from nature, often recklessly, but didn't give back. Some even turned parts of the world into deserts by accident, like the Sahara and the Chihuahua. They were unable to think about how their actions would affect your life because they could only look at one thing at a time when they made decisions. Thank God people have evolved to manage complex things like nature and economies holistically. Now you can leave the world better than you found it. I hope, by your time, deserts have become green and productive again and keep filling up with abundant life.

Many people used to read books they called holy books. The first story in the Bible was about creation and the original sin of humankind. The story talked about the way we humans often kill animals the Creator had made and thought were so good.
My generation still does this. I feel bad about this and used to ask God to forgive me, but it is really you I need to ask for forgiveness. I hope your generation evolves to finally live up to the meaning of *Homo sapiens*: "wise human."

That is the best hope for your grandchildren.

Love,

*Great-gran Siobhán*

**Great-grandmother Siobhán**

# Bibliography

Aburto-Oropeza, Octavio, Brad Erisman, Grantly R. Galland, Ismael Mascareñas-Osorio, Enric Sala, and Exequiel Ezcurra. "Large Recovery of Fish Biomass in a No-Take Marine Reserve." *PLOS ONE* 6, no. 8 (2011): e23601. https://doi.org/10.1371/journal.pone.0023601.

Aburto-Oropeza, Octavio, Valentina Platzgummer, Erica M. Ferrer, Catalina López-Sagástegui, Rocío de Guadalupe Abud Mirabentet et al. "Marine Prosperity Areas : A Framework for Aligning Ecological Restoration and Human Well-Being Using Area-Based Protections." *Frontiers in Marine Science* 11 (2025): 1–16. https://doi.org/10.3389/fmars.2024.1491483.

Alkonis, Steve and Derek Kerber. "Lahaina Fire Incident Analysis Report." Fire Safety Research Institute (FSRI), September 14, 2024. https://doi.org/10.60752/102376.26858962.

American Farm Bureau. "American Farm Bureau Convention Monday Recap: Remarks from Ag Secretary Vilsack, Ag Innovation Challenge and YF&R Awards, Farm Bill Congressional Discussion and More." News release, January 22, 2024.

Amsili, Joseph, Harold van Es, Robert Schindelbeck, et al. *Characterization of Soil Health in New York State: Technical Report.* New York Soil Health Initiative, Cornell University, 2020. https://doi.org/10.13140/ RG.2.2.17826.30405.

Andermann, Tobias, Søren Faurby, Samuel T. Turvey, Alexandre Antonelli, and Daniele Silvestro. "The Past and Future Human Impact on Mammalian Diversity." *Science Advances* 6, no. 36 (2020). https:// doi.org/10.1126/sciadv.abb2313.

Archer Daniels Midland. "ADM Reports Fourth Quarter and Full-Year 2024 Results; Provides 2025 Guidance," February 4, 2025. https://investors.adm.com/news/news-details/2025/ADM-Reports-Fourth-Quarter-and-Full-Year-2024-Results-Provides-2025-Guidance/

Arslanov, Kh. A., G. T. Cook, Steinar Guliksen, et al. "Consensus Dating of Mammoth Remains from Wrangel Island." *Radiocarbon* 40, no. 1 (2006): 289–94. https://doi.org/10.1017/S0033822200018166.

Audubon, John James. *Ornithological Biography.* Adam and Charles Black, 1831.

Balesdent, Jérôme, Isabelle Basile-Doelsch, Joël Chadoeuf, et al. "Atmosphere–Soil Carbon Transfer as a Function of Soil Depth." *Nature* 559, no. 7715 (2018): 599–602. https://doi.org/10.1038/s41586-018-0328-3.

Barnosky, Anthony D. "Megafauna Biomass Tradeoff as a Driver of Quaternary and Future Extinctions." *Proceedings of the National Academy of Sciences of the United States of America* 105 (supplement 1) (2008): 11543–48. https://doi.org/10.1073/pnas.0801918105.

Bar-On, Yinon M., Rob Phillips, and Ron Milo. "The Biomass Distribution on Earth." *Proceedings of the National Academy of Sciences of the United States of America* 115, no. 25 (2018): 6506–11. https://doi. org/10.1073/pnas.1711842115.

Barros, Beatriz, and Richard Wilk. "The Outsized Carbon Footprints of the Super-Rich." *Sustainability: Science, Practice and Policy* 17, no. 1 (2021): 316–22. https://doi.org/10.1080/15487733.2021.1949847.

Bauska, Thomas K, "Ice Core Records of Atmospheric Carbon Dioxide." *In Encyclopedia of Quaternary Science (Third edition)*, edited by Scott Elias, 122-137. Elsevier, 2025. ISBN 9780443299971. https://doi.org/10.1016/B978-0-323-99931-1.00264-6.

Bauska, Thomas K., Daniel Baggenstos, Edward J. Brook, et al. "Carbon Isotopes Characterize Rapid Changes in Atmospheric Carbon Dioxide During the Last Deglaciation." *PNAS* 113, no. 13 (2016): 3465–70, https://doi.org/10.1073/pnas.1513868113.

Bekele, Wondimagegne, Abdulai Guinguina, Abiy Zegeye, Addis Simachew, and Mohammad Ramin. "Contemporary Methods of Measuring and Estimating Methane Emission from Ruminants" *Methane* 1, no. 2 (2022): 82–95. https://doi.org/10.3390/methane1020008.

Bergman, Juraj, Rasmus Ø. Pedersen, Erick J. Lundgren, et al. "Worldwide Late Pleistocene and Early Holocene Population Declines in Extant Megafauna Are Associated with *Homo sapiens* Expansion Rather than Climate Change." *Nature Communications* 14:7679, no. 1 (2023). https://doi.org/10.1038/ s41467-023-43426-5.

Bianchi, Daniele, David A. Carozza, Eric D. Galbraith, Jérôme Guiet, and Timothy DeVries. "Estimating Global Biomass and Biogeochemical Cycling of Marine Fish with and Without Fishing." *Science Advances* 7, no. 41 (2021). https://doi.org/10.1126/sciadv.abd7554.

Black, Simon, Antung A. Liu, Ian W.H. Parry, and Nate Vernon. "IMF Fossil Fuel Subsidies Data: 2023 Update." IMF Working Papers 2023, 169, (2023). https://doi.org/10.5089/9798400249006.001.

Borrelli, Pasquale, David A. Robinson, Larissa R. Fleischer, et al. "An Assessment of the Global Impact of 21st Century Land Use Change on Soil Erosion." *Nature Communications* 8, no. 1 (2017): 2013. https:// doi.org/10.1038/s41467-017-02142-7.

Bowles, Samuel, and Henry T. Wright. "Cultivation of Cereals by the First Farmers Was Not More Productive than Foraging." *Proceedings of the National Academy of Sciences of the United States of America* 108, no. 12 (2011): 4760–65. https://doi.org/10.1073/pnas.1010733108.

Boyce, Daniel G., Marlon R. Lewis, and Boris Worm. "Global Phytoplankton Decline over the Past Century." *Nature* 466, no. 7306 (2010): 591–96. https://doi.org/10.1038/nature09268.

Brandt, Jessica E., Jeff S. Wesner, Gregory T. Ruggerone, et al. "Continental-Scale Nutrient and Contaminant Delivery by Pacific Salmon." *Nature* 634, no. 8035 (2024): 875–82. https://doi.org/10.1038/s41586-024-07980-2.

Bråthen, Kari Anne, Francisco I. Pugnaire, and Richard D. Bardgett. "The Paradox of Forbs in Grasslands and the Legacy of the Mammoth Steppe." *Frontiers in Ecology and the Environment* 19, no. 10 (2021): 584–92. https://doi.org/10.1002/fee.2405.

Butterfield, Jody, Sam Bingham, and Allan Savory. *Holistic Management Handbook: Regenerating Your Land and Growing Your Profits*. 3rd ed. Island Press, 2019. https://doi.org/10.5822/978-1-61091-977-7.

Buzhdygan, Oksana Y., Sebastian T. Meyer, Wolfgang W. Weisser, et al. "Biodiversity Increases Multitrophic Energy Use Efficiency, Flow and Storage in Grasslands." *Nature Ecology & Evolution* 4, no. 3 (2020): 393–405. https://doi.org/10.1038/s41559-020-1123-8.

Byrne, Brendan, Junjie Liu, Kevin W. Bowman, et al. "Carbon Emissions from the 2023 Canadian Wildfires." *Nature* 633, no. 8031 (2024): 835–39. https://doi.org/10.1038/s41586-024-07878-z.

Bysouth, David, Julee J. Boan, Jay R. Malcolm, and Anthony R. Taylor. "High Emissions or Carbon Neutral? Inclusion of 'Anthropogenic' Forest Sinks Leads to Underreporting of Forestry Emissions." *Frontiers in Forests and Global Change* 6 (2024). https://doi.org/10.3389/ffgc.2023.1297301.

Cain, Zachary, and Stephen Lovejoy. "History and Outlook for Farm Bill Conservation Programs." *Choices* 19, no. 4 (2004): 37–42. https://www.choicesmagazine.org/2004-4/policy/2004-4-09.htm.

California Department of Pesticide Regulation (CDPR). *Pesticide Use Reporting—2023 Summary Data, Carcinogens: Pounds Applied.* 4. https://www.cdpr.ca.gov/wp-content/uploads/2025/08/2023_ carcinogens_pounds_applied.pdf.

Cannon, Scott, "Water on Fire from Fracking Gas Drilling." posted by Gas Drilling Awareness Coalition.Accessed January 15, 2026. https://youtu.be/4LBjSXWQRV8?si=CCxSNN4qJnNNu7XG.

Cassidy, Lisa R., and Gary Kleppel. "The Effect of Grazing Regime on Grassland Bird Abundance in New York State." *Northeastern Naturalist* 24, no. sp8 (2017): 86–98. https://doi.org/10.1656/045.024.0sp807.

Chambers, Jeffrey Q., Robinson I. Negron-Juarez, Daniel Magnabosco Marra, et al. "The Steady-State Mosaic of Disturbance and Succession across an Old-Growth Central Amazon Forest Landscape." *Proceedings of the National Academy of Sciences of the United States of America* 110, no. 10 (2013): 3949–54. https://doi.org/10.1073/pnas.1202894110.

Chandan, Deepak, and W. Richard Peltier. "African Humid Period Precipitation Sustained by Robust Vegetation, Soil, and Lake Feedbacks." *Geophysical Research Letters* 47, no. 21 (2020): e2020GL088728. https://doi.org/10.1029/2020GL088728.

Chatters, James C., Ben A. Potter, Stuart J. Fiedel, Juliet E. Morrow, Christopher N. Jass, and Matthew J. Wooller. "Mammoth Featured Heavily in Western Clovis Diet." *Science Advances* 10, no. 49 (2024): eadr3814. https://doi.org/10.1126/sciadv.adr3814.

Chaurasia,, Aalok. "Population Effects of Increase in World Energy Use and $CO_2$ Emissions: 1990–2019." *Journal of Population and Sustainability* 5, no. 1 (2020): 87–125. http://doi.org/10.3197/jps.2020.5.1.87.

Chellemi, D. O. "Plant Health Management: Soil Fumigation." *Encyclopedia of Agriculture and Food Systems* 1 (2014): 456–59. https://doi.org/10.1016/B978-0-444-52512-3.00250-3.

Christensen, Martin-Brehm, Christian Hallum, Alex Maitland, et al. *Survival of the Richest: How We Must Tax the Super-Rich Now to Fight Inequality.* Oxfam briefing paper, 2023. https://financing.desa. un.org/sites/default/files/2023-03/OXFAM_Input%20Tax%20Report.pdf

Clarkson, Chris, Zenobia Jacobs, Ben Marwick, et al. "Human Occupation of Northern Australia by 65,000 Years Ago." *Nature* 547, no. 7663 (2017): 306–10. https://doi.org/10.1038/nature22968.

Clover, Charles. *Rewilding the Sea: How to Save Our Oceans.* Witness Books, 2022.

Cohen, Edward E. "Inequality." *In Roman Inequality,* 26–48. Oxford University Press, 2023. https://doi. org/10.1093/oso/9780197687345.003.0002.

Cohen, Tim J., John D. Jansen, Luke A. Gliganic, et al. "Hydrological Transformation Coincided with Megafaunal Extinction in Central Australia." *Geology* 43, no. 3 (2015): 195–98. https://doi.org/10.1130/ G36346.1.

Conant, Richard T., Johan Six, and Keith Paustian. "Land Use Effects on Soil Carbon Fractions in the Southeastern United States. I. Management-Intensive Versus Extensive Grazing." *Biology and Fertility of Soils* 38, no. 6 (2003): 386–92. https://doi. org/10.1007/s00374-003-0652-z.

Cordain, L., S. B. Eaton, J. Brand Miller, N. Mann, and K. Hill. "The Paradoxical Nature of Hunter-Gatherer Diets: Meat-Based, yet Non-Atherogenic." *European Journal of Clinical Nutrition* 56, S1 (2002): S42–52. https://doi.org/10.1038/sj.ejcn.1601353.

Coulson, David and Alec Campbell. "Rock Art of the Tassili n Ajjer, Algeria." *Adoranten* (2010): 25–38, fig. 18. https://www. rockartscandinavia.com/images/articles/ coulsona10.pdf.

Coulthard, Sally. *A Short History of The World According to Sheep.* Bloomsbury Publishing, 2021.

Crandall, Leslie. *Soil Survey of Chenango County, New York.* USDA Soil Conservation Service 1985.

Cunningham, Calum X., Grant J. Williamson, and David M. J. S. Bowman. "Increasing Frequency and Intensity of the Most Extreme Wildfires on Earth." *Nature Ecology & Evolution* 8, no. 8 (2024): 1420–25. https://doi.org/10.1038/s41559-024-02452-2.

Dass, Pawlok, Benjamin Z Houlton, Yingping Wang, and David Warlind. "Grasslands May Be More Reliable Carbon Sinks than Forests in California." *Environmental Research Letters* 13, no. 7 (2018): 074027. https://doi.org/10.1088/1748-9326/aacb39.

DeConto, Robert M., Simone Galeotti, Mark Pagani, et al. "Past Extreme Warming Events Linked to Massive Carbon Release from Thawing Permafrost." *Nature* 484, no. 7392 (2012): 87–91. https://doi. org/10.1038/nature10929.

Demarchi, Beatrice, Josefin Stiller, Alicia Grealy, et al. "Ancient Proteins Resolve Controversy over the Identity of *Genyornis* Eggshell." *Proceedings of the National Academy of Sciences of the United States of America* 119, no. 43 (2022); e2109326119. https://doi.org/10.1073/pnas.2109326119.

Dembitzer, Jacob, Ran Barkai, Miki Ben-Dor, and Shai Meiri. "Levantine Overkill: 1.5 Million Years of Hunting Down the Body Size Distribution." *Quaternary Science Reviews* 276 (2022): 107316. https:// doi.org/10.1016/j.quascirev.2021.107316.

den Hartigh, Laura J. "Conjugated Linoleic Acid Effects on Cancer, Obesity, and Atherosclerosis: A Review of Pre-Clinical and Human Trials with Current Perspectives." *Nutrients* 11, no. 2 (2019): 370. https:// doi.org/10.3390/nu11020370.

DeSilva, Jeremy M., James F. A. Traniello, Alexander G. Claxton, and Luke D. Fannin. "When and Why Did Human Brains Decrease in Size? A New Change-Point Analysis and Insights From Brain Evolution in Ants." *Frontiers in Ecology and Evolution* 9 (2021). https://doi.org/10.3389/fevo.2021.742639.

Doughty, Christopher E., Adam Wolf, and Christopher B. Field. "Biophysical Feedbacks Between the Pleistocene Megafauna Extinction and Climate: The First Human-Induced Global Warming?" *Geophysical Research Letters* 37, no. 15 (2010): L15703. https://doi.org/10.1029/2010GL043985.

Doughty, Christopher E., Adam Wolf, and Yadvinder Malhi. "The Legacy of the Pleistocene Megafauna Extinctions on Nutrient Availability in Amazonia." *Nature Geoscience* 6, no. 9 (2013): 761–64. https:// doi.org/10.1038/ngeo1895.

Dunne, John P., Robert A. Armstrong, Anand Gnanadesikan, and Jorge L. Sarmiento. "Empirical and Mechanistic Models for the Particle Export Ratio." *Global Biogeochemical Cycles* 19 (2005): GB4026. https://doi.org/10.1029/2004GB002390.

Eldridge, David J., Fernando T. Maestre, Terry B. Koen, and Manuel Delgado-Baquerizo. "Australian Dryland Soils Are Acidic and Nutrient-Depleted, and Have Unique Microbial Communities Compared with Other Drylands." *Journal of Biogeography* 45, no. 12 (2018): 2803–14. https://doi.org/10.1111/ jbi.13456.

Epplett, William Christopher. "*Animal Spectacula of the Roman Empire.*" PhD diss., University of British Columbia, 2001. https://open.library.ubc.ca/soa/cIRcle/collections/ubctheses/831/items/1.0090535

Estes, James A., John Terborgh, Justin S. Brashares, et al. "Trophic Downgrading of Planet Earth." *Science* 333, no. 6040 (2011): 301–6. https://doi.org/10.1126/science.1205106.

Fairfax, Emily and Cherie Westbrook. "The Ecology and Evolution of Beavers: Ecosystem Engineers That Ameliorate Climate Change." *Annual Review Ecology, Evolution, and Systematics* 55 (2024): 323–45. https://doi.org/10.1146/annurev-ecolsys-102722-122317.

Farr, Jonathan James, and Clifford A. White."Buffalo on the Edge: Factors Affecting Historical Distribution and Restoration of Bison Bison in the Western Cordillera, North America." *Diversity* 14, no. 11 (2022): 937. https://doi.org/10.3390/d14110937.

Feng, Yu, Zhenzhong Zeng, Timothy D. Searchinger, et al. "Doubling of Annual Forest Carbon Loss over the Tropics during the Early Twenty-First Century." *Nature Sustainability* 5, no. 5 (2022): 444–51. https://doi.org/10.1038/s41893-022-00854-3

Florida Museum. *Yearly Worldwide Shark Attack Summary: The ISAF 2025 Shark Attack Report.* https:// www.floridamuseum.ufl.edu/shark-attacks/yearly-worldwide-summary/

Food and Agriculture Organization of the United Nations. *Report of the World Food Summit.* Food and Agriculture Organization of the United Nations, 1996, https://www.fao.org/4/w3548e/w3548e00. htm#Boutros.

Fordham, Damien A., Stuart C. Brown, H. Reşit Akçakaya, et al. "Process-Explicit Models Reveal Pathway to Extinction for Woolly Mammoth Using Pattern-Oriented Validation." *Ecology Letters* 25, no. 1 (2022): 125–37. https://doi.org/10.1111/ele.13911.

Foster, David R. "Land-Use History (1730–1990) and Vegetation Dynamics in Central New England, USA." *Journal of Ecology* 80, no. 4 (1992): 753. https://doi.org/10.2307/2260864.

Francis. Laudato Si': On Care for Our Common Home. Vatican Press, 2015. https://www.vatican.va/content/ francesco/en/encyclicals/documents/papa-francesco_20150524_enciclica-laudato-si.html.

Frerick, Austin. *Barons: Money, Power, and the Corruption of America's Food Industry.* Island Press, 2024.

Gauci, Vincent, Sunitha Rao Pangala, Alexander Shenkin, et al. "Global Atmospheric Methane Uptake by Upland Tree Woody Surfaces." *Nature* 631, no. 8022 (2024): 796–800. https://doi.org/10.1038/s41586-024-07592-w.

Ghoneim, E., and F. El-Baz. "DEM-Optical-Radar Data Integration for Palaeohydrological Mapping in the Northern Darfur, Sudan: Implication for Groundwater Exploration." *International Journal of Remote Sensing* 28, no. 22 (2007): 5001–18. https://doi.org/10.1080/01431160701266818.

Gilens, Martin, and Benjamin I. Page. "Testing Theories of American Politics: Elites, Interest Groups, and Average Citizens." *Perspectives on Politics* 12, no. 3 (2014): 564–81. https://doi.org/10.1017/ S1537592714001595.

Gill, Jacquelyn L., John W. Williams, Stephen T. Jackson, Katherine B. Lininger, and Guy S. Robinson. "Pleistocene Megafaunal Collapse, Novel Plant Communities, and Enhanced Fire Regimes in North America." *Science* 326, no. 5956 (2009): 1100–1103. https://doi.org/10.1126/science.1179504.

Gornitz, Vivien, "The Great Ice Meltdown and Rising Seas: Lessons for Tomorrow." NASA Goddard Institute of Space (2012). Updated April 3, 2025. Accessed July 26, 2025. https://sites.law.lsu.edu/coast/2021/01/nasa-science-briefs-the-great-ice-meltdown-and-rising-seas-lessons-for-tomorrow/

Greenspoon, Lior, Eyal Krieger, Ron Sender, et al. "The Global Biomass of Wild Mammals." *Proceedings of the National Academy of Sciences of the United States of America* 120, no. 10 (2023). https://doi. org/10.1073/pnas.2204892120.

Guagnin, Maria, Angela R. Perri, and Michael D. Petraglia. "Pre-Neolithic Evidence for Dog-Assisted Hunting Strategies in Arabia." *Journal of Anthropological Archaeology* 49 (2018): 225–236, ISSN 0278-4165, https://doi.org/10.1016/j.jaa.2017.10.003.

Haddaway, Neal R., Katarina Hedlund, Louise E. Jackson, et al. "How Does Tillage Intensity Affect Soil Organic Carbon? A Systematic Review." *Environmental Evidence* 6, no. 1 (2017): 30. https://doi. org/10.1186/s13750-017-0108-9.

Heinberg, Richard, and Michael Bomford, *The Food and Farming Transition: Toward a Post-Carbon Food System.* Post Carbon Institute, 2009.

Helmig, D., C. R. Thompson, J. Evans, P. Boylan, J. Hueber, and J.-H. Park. "Highly Elevated Atmospheric Levels of Volatile Organic Compounds in the Uintah Basin, Utah." *Environmental Science & Technology* 48, no. 9 (2014): 4707–15. https://doi.org/10.1021/es405046r.

Henneberg, M., and M. Steyn. "Trends in Cranial Capacity and Cranial Index in Subsaharan Africa During the Holocene." *American Journal of Human Biology* 5, no. 4 (1993): 473–79. https://doi.org/10.1002/ ajhb.1310050411.

Holdaway, Richard N., Morten E. Allentoft, Christopher Jacomb, Charlotte L. Oskam, Nancy R. Beavan, and Michael Bunce. "An Extremely Low-Density Human Population Exterminated New Zealand Moa." *Nature Communications* 5, no. 1 (2014): 5436. https://doi.org/10.1038/ncomms6436.

Holdo, Ricardo M., Anthony R. E. Sinclair, Andrew P. Dobson, et al. "A Disease-Mediated Trophic Cascade in the Serengeti and Its Implications for Ecosystem C." *PLOS Biology* 7, no. 9 (2009): e1000210. https:// doi.org/10.1371/journal.pbio.1000210.

Hooijer, A., and R. Vernimmen. "Global LiDAR Land Elevation Data Reveal Greatest Sea-Level Rise Vulnerability in the Tropics." *Nature Communications* 12, no. 1 (2021): 3592. https://doi.org/10.1038/ s41467-021-23810-9.

Houghton, R. A. "Balancing the Global Carbon Budget." *Annual Review of Earth and Planetary Sciences* 35, no. 1 (2007): 313–47. https://doi.org/10.1146/annurev.earth.35.031306.140057.

Houston, Jordon Alex. 2024. "Exotics for Entertainment: A Reconstruction of the Roman Exotic Beast Trade (First to Third Centuries AD)". *Theoretical Roman Archaeology Journal* 7(1):1–39. Archaeozoological Evidence. DOI: https://doi.org/10.16995/traj.10158.

Intergovernmental Science-Policy Platform on Biodiversity and Ecosystem Services (IPBES). *Summary for Policymakers of the Global Assessment Report on Biodiversity and Ecosystem Services of the Intergovernmental Science-Policy Platform on Biodiversity and Ecosystem Services.* IPBES, 2019. https://files.ipbes.net/ipbes-web-prod-public-files/inline/files/ipbes_global_assessment_report_ summary_for_policymakers.pdf.

IPC (Integrated Food Security Phase Classification). *Sudan: Acute Food Insecurity Snapshot: April 2024–February 2025.* June 27, 2024. https://www.ipcinfo.org/fileadmin/user_upload/ipcinfo/docs/ IPC_Sudan_Acute_Food_Insecurity_Snapshot_Jun2024_Feb2025.pdf.

IPC (Integrated Food Security Phase Classification). *Sudan: IPC Acute Food Insecurity Snapshot: June 2023–February 2024.* August 2, 2023. https://www.ipcinfo.org/fileadmin/user_upload/ipcinfo/docs/ IPC_Sudan_Acute_Food_Insecurity_Jun2023_Feb2024_snapshot.pdf.

Iriarte, José, Michael J. Ziegler, Alan K. Outram, et al. "Ice Age Megafauna Rock Art in the Colombian Amazon?" *Philosophical Transactions of the Royal Society B: Biological Sciences* 377, no. 1849 (2022). https://doi.org/10.1098/rstb.2020.0496.

Jackson, Randall D. "Grazed Perennial Grasslands Can Match Current Beef Production While Contributing to Climate Mitigation and Adaptation," *Agricultural & Environmental Letters* 7: e20059 (2022). https:// doi.org/10.1002/ael2.20059.

Janssen, Marco A., and Marten Scheffer. "Overexploitation of Renewable Resources by Ancient Societies and the Role of Sunk-Cost Effects." *Ecology and Society* 9, no. 1 (2004): 6. http://www. ecologyandsociety.org/vol9/iss1/art6.

Jasechko, Scott, Hansjörg Seybold, Debra Perrone, et al. "Rapid Groundwater Decline and Some Cases of Recovery in Aquifers Globally." *Nature* 625, no. 7996 (2024): 715–21. https://doi.org/10.1038/s41586-023-06879-8.

Jehne, Walter. "Supporting the Soil Carbon Sponge." Interview by Tracy Frisch. *EcoFarming Daily,* June 20, 2019.

Jehne, Walter. "The Soil Carbon Sponge, Climate Solutions and Healthy Water Cycles with Walter Jehne." Recorded by Biodiversity for a Livable Climate. Presented on April 26, 2018 at Harvard University. https://youtu.be/123y7jDdbfY?si=v8f9ToXnf6AZTC6-.

Jehne, Walter. "Restoring Water Cycles to Naturally Cool Climate." Remarks to the Soil Carbon Sponge seminar, November 30, 2016. Video, 2 hr., 15 min., 24 sec. https://youtu.be/K4ygsdHJjdI.

John, Daisy A., and Giridhara R. Babu. "Lessons From the Aftermaths of Green Revolution on Food System and Health." *Frontiers in Sustainable Food Systems* 5 (2021). https://doi.org/10.3389/fsufs.2021.644559.

Johnson, Chris N., Susan Rule, Simon G. Haberle, A. Peter Kershaw, G. Merna McKenzie, and Barry W. Brook. "Geographic Variation in the Ecological Effects of Extinction of Australia's Pleistocene Megafauna." *Ecography* 39, no. 2 (2016): 109–16. https://doi.org/10.1111/ecog.01612.

Johnson, Christopher N., Lynda D. Prior, Sally Archibald, et al. "Can Trophic Rewilding Reduce the Impact of Fire in a More Flammable World?" *Philosophical Transactions of the Royal Society B: Biological Sciences* 373, no. 1761 (2018): 20170443. https://doi.org/10.1098/rstb.2017.0443.

Johnson, Eileen, and Leland C. Bement. "Bison Butchery at Cooper, a Folsom Site on the Southern Plains." *Journal of Archaeological Science* 36, no. 7 (2009): 1430–46. https://doi.org/10.1016/j.jas.2009.02.007.

Jones, D. L., C. Nguyen, and R. D. Finlay. "Carbon Flow in the Rhizosphere: Carbon Trading at the Soil–Root Interface." *Plant and Soil* 321, no. 1–2 (2009): 5–33. https://doi.org/10.1007/s11104-009-9925-0.

Jones, Holly.P., Jennifer Appoo, Cassandra.E. Benkwitt, et al. "The Circular Seabird Economy is Critical for Oceans, Islands and People." *Nature Review Biodiversity* 1, (2025): 689–702. https://doi.org/10.1038/s44358-025-00099-w

Jones, Harold. 2017. *Audubon Gallery Catalog.* Manlius Library, Syracuse University, 2017.

Jones, Kendall R., Oscar Venter, Richard A. Fuller, et al. "One-Third of Global Protected Land Is Under Intense Human Pressure." *Science* 360, no. 6390 (2018): 788–91. https://doi.org/10.1126/science. aap9565.

Kaars, Sander van der, Gifford H. Miller, Chris S. M. Turney, et al. "Humans Rather than Climate the Primary Cause of Pleistocene Megafaunal Extinction in Australia." *Nature Communications* 8 (2017): 14142. https://doi.org/10.1038/ncomms14142.

Kaplan, Thomas. "Citing Health Risks, Cuomo Bans Fracking in New York State." *The New York Times,*Dec. 17, 2014. https://www.nytimes.com/2014/12/18/nyregion/cuomo-to-banfracking-in-new-york-state-citing-health-risks.html.

Karp, Allison T., J. Tyler Faith, Jennifer R. Marlon, and A. Carla Staver. "Global Response of Fire Activity to Late Quaternary Grazer Extinctions." *Science* 374, no. 6571 (2021): 1145–48. https://doi.org/10.1126/ science.abj1580.

Kerber, Steve, and Derek Alkonis. *Lahaina Fire Incident Analysis Report.* Fire Safety Research Institute.September 14, 2024. https://doi.org/10.60752/102376.26858962.

Kim, Jungho. "Women's Education and Fertility: An Analysis of the Relationship between Education and Birth Spacing in Indonesia." *Economic Development and Cultural Change* 58, no. 4 (2010): 739–74. https://doi.org/10.1086/649638.

Kleiber, Max. "Body Size and Metabolic Rate." *Physiological Reviews* 27, no. 4 (1947): 511–41. https://doi. org/10.1152/physrev.1947.27.4.511.

Kleppel, Gary S. "Do Differences in Livestock Management Practices Influence Environmental Impacts?" *Frontiers in Sustainable Food Systems.* 4 (2020). https://doi.org/10.3389/fsufs.2020.00141.

Kristensen, Jeppe A., Jens-Christian Svenning, Katerina Georgiou, and Yadvinder Malhi. "Can Large Herbivores Enhance Ecosystem Carbon Persistence?" *Trends in Ecology & Evolution* 37, no. 2 (2022): 117–28. https://doi.org/10.1016/j.tree.2021.09.006.

Lal, Rattan. "Conceptual Basis of Managing Soil Carbon: Inspired by Nature and Driven by Science." *Journal of Soil and Water Conservation* 74, no. 2 (2019): 29A–34A. https://doi.org/10.2489/jswc.74.2.29A.

Lal, Rattan, Pete Smith, Hermann F. Jungkunst, et al. "The Carbon Sequestration Potential of Terrestrial Ecosystems." *Journal of Soil and Water Conservation* 73, no. 6 (2018): 145A–152A. https://doi. org/10.2489/jswc.73.6.145A.

Lalampaa, Priscilla K., Oliver V. Wasonga, Daniel I. Rubenstein, and Jesse T. Njoka. "Effects of Holistic Grazing Management on Milk Production, Weight Gain, and Visitation to Grazing Areas by Livestock and Wildlife in Laikipia County, Kenya." *Ecological Processes* 5, no. 17, (2016). https://doi.org/10.1186/ s13717-016-0061-5.

Lamperty, Therese, Kai Zhu, John R. Poulsen, and Amy E. Dunham. "Defaunation of Large Mammals Alters Understory Vegetation and Functional Importance of Invertebrates in an Afrotropical Forest." *Biological Conservation* 241 (2020): 108329. https://doi. org/10.1016/j.biocon.2019.108329.

Larrasoaña, Juan C., Andrew P. Roberts, and Eelco J. Rohling. "Dynamics of Green Sahara Periods and Their Role in Hominin Evolution." *PLOS ONE* 8, no. 10 (2013): e76514. https://doi.org/10.1371/journal. pone.0076514.

Larson, E, C. Greig, J. Jenkins, et al. *Net-Zero America: Potential Pathways, Infrastructure, and Impacts, Final Report Summary.* Princeton University, 2021. https://netzeroamerica.princeton.edu/img/ Princeton%20NZA%20FINAL%20REPORT%20SUMMARY%20(29Oct2021).pdf

Lee, Richard Borshay. T*he !Kung San: Men, Women, and Work in a Foraging Society,* 250–80. Cambridge University Press, 1979.

Lee, YC, Chang KY, Sethi S. "Association of Chronic Lower Respiratory Disease With County Health Disparities in New York State." *JAMA, Net Open* 4, no. 11 (2021): doi: 10.1001/ jamanetworkopen.2021.34268.

Liang, Xiao, Nan Tian, Diego Lopes da Silva, Lorenzo Scarazzato, Zubaida A. Karim, and Jade Guiberteau Ricard. *Trends in World Military Expenditure,* 2024. Stockholm International Peace Research Institute, 2025. https://doi.org/10.55163/AVEC8366.

Lin, Yucheng, Fiona D. Hibbert, Pippa L. Whitehouse, et al. "A Reconciled Solution of Meltwater Pulse 1A Sources Using Sea-Level Fingerprinting." *Nature Communications* 12, no. 1 (2021): 2015. https://doi. org/10.1038/s41467-021-21990-y.

Lutz, Michael J., Ken Caldeira, Robert B. Dunbar, and Michael J. Behrenfeld. "Seasonal Rhythms of Net Primary Production and Particulate Organic Carbon Flux to Depth Describe the Efficiency of Biological Pump in the Global Ocean." *Journal of Geophysical Research* 112, no. C10 (2007): C10011. https://doi.org/10.1029/2006JC003706.

Maitland, Alex, Max Lawson, Hilde Stroot, Alexandre Poidatz, Ashfaq Khalfan, and Nafkote Dabi. *Carbon Billionaires: The Investment Emissions of the World's Richest People,* Oxfam briefing note, 2022. https:// doi.org/10.21201/2022.9684.

Makarieva, Anastassia M., and Victor G. Gorshkov. "The Biotic Pump: Condensation, Atmospheric Dynamics and Climate." *International Journal of Water 5,* no. 4 (2010): 365–385. https://doi. org/10.1504/IJW.2010.038729.

Malhi, Yadvinder, Christopher E. Doughty, Mauro Galetti, Felisa A. Smith, Jens-Christian Svenning, and John W. Terborgh. "Megafauna and Ecosystem Function from the Pleistocene to the Anthropocene." *Proceedings of the National Academy of Sciences of the United States of America* 113, no. 4 (2016): 838–46. https://doi.org/10.1073/pnas.1502540113.

Mariani, Gaël, William W. L. Cheung, Arnaud Lyet, et al. "Let More Big Fish Sink: Fisheries Prevent Blue Carbon Sequestration—Half in Unprofitable Areas." *Science Advances* 6, no. 44 (2020). https://doi. org/10.1126/sciadv.abb4848.

Markgraf, Vera. "Younger Dryas in Southern South America?," *Boreas* 20, no. 1 (2008): 63–69. https://doi. org/10.1111/j.1502-3885.1991.tb00459.x.

McCrummen, Stephanie. "'This Will Finish Us': How Gulf Princes, The Safari Industry and Conservation Groups Are Displacing the Maasai From the Last of Their Serengeti Homeland." *The Atlantic*, April 8, 2024.

Menking, James A., Sarah A. Shackleton, Thomas K. Bauska, et al. "Multiple Carbon Cycle Mechanisms Associated with the Glaciation of Marine Isotope Stage 4." *Nature Communications* 13, no. 1 (2022): 5443. https://doi.org/10.1038/s41467-022-33166-3.

Merz, Joseph J., Phoebe Barnard, William E. Rees, et al. "World Scientists' Warning: The Behavioural Crisis Driving Ecological Overshoot." *Science Progress* 106, no. 3 (2023). https://doi. org/10.1177/00368504231201372.

Miller, Gifford H., Marilyn L. Fogel, John W. Magee, Michael K. Gagan, Simon J. Clarke, and Beverly J. Johnson. "Ecosystem Collapse in Pleistocene Australia and a Human Role in Megafaunal Extinction." Science 309, no. 5732 (2005): 287–90. https://doi.org/10.1126/science.1111288.

Miller, Michelle. "A Breakdown of the Food Dollar: Where Does the Money Go?" *AGDAILY*, July 23, 2024. https://www.agdaily.com/insights/food-dollar-where-does-the-money-go/.

Mitsunaga, Bryce A., Amy M. Jewell, Solana Buchanan, et al. "Fundamentally Unchanged Northwestern African Rainfall Regimes across the Plio-Pleistocene Transition." *Science Advances* 11, no. 25 (2025). https://doi.org/10.1126/sciadv.ads3149.

Monique Ross and Ben Cheshire. "Meet the Defiant Grazier Using Feral Donkeys to Regenerate His Land at Kachana Station in an 'Illegal' Experiment." ABC News, November 3, 2024. https:// www.abc.net.au/news/2024-11-04/grazier-using-wild-donkeys-to-regenerate-land-in-legal-battle/104445766?future=true&.

Montgomery, David R., Anne Biklé, Ray Archuleta, Paul Brown, and Jazmin Jordan. "Soil Health and Nutrient Density: Preliminary Comparison of Regenerative and Conventional Farming." *PeerJ* 10 (2022): e12848. https://doi.org/10.7717/peerj.12848.

Morcote-Ríos, Gaspar, Francisco Javier Aceituno, José Iriarte, Mark Robinson, and Jeison L. Chaparro-Cárdenas. "Colonisation and Early Peopling of the Colombian Amazon During the Late Pleistocene and the Early Holocene: New Evidence from La Serranía La Lindosa." *Quaternary International* 578 (2021): 5–19. https://doi.org/10.1016/j.quaint.2020.04.026.

Morrison, Steve, Tony Azevedo, and Darlene Coehoorn to Chuck Connor, September 25, 2007. https:// nodpa.com/files/2007-09-25_Letter_to_Acting_Secretary_Conner.pdf.

Muir, John. *My First Summer in the Sierra*. Sierra Club Books, 1988. Originally published in 1911 by Houghton Mifflin.

Murchie, Tyler J., Alistair J. Monteath, Matthew E. Mahony, et al. "Collapse of the Mammoth-Steppe in Central Yukon as Revealed by Ancient Environmental DNA." *Nature Communications* 12, no. 1 (2021): 7120. https://doi.org/10.1038/s41467-021-27439-6.

Naidu, Dilip G. T., Shamik Roy, and Sumanta Bagchi. "Loss of Grazing by Large Mammalian Herbivores Can Destabilize the Soil Carbon Pool." *Proceedings of the National Academy of Sciences of the United States of America* 119, no. 43 (2022): e2211317119. https://doi.org/10.1073/pnas.2211317119.

National Agricultural Statistics Service. *2024 State Agriculture Overview: California* Accessed August 7, 2025. https://www.nass.usda.gov/Quick_Stats/Ag_Overview/stateOverview.php?state=CALIFORNIA#:~:text=USDA/NASS%202024%20State%20Agriculture,40%2C283%2C000%2C000.

New York Academy of Sciences. "Responding to Climate Change in New York State: The Clim AID Integrated Assessment for Effective Climate Change Adaptation in New York State." *Annals of the New York Academy of Sciences* 1244, no. 1 (2011): 2–649. https://doi.org/10.1111/j.1749-6632.2011.06331.x.

Niklaus, Pascal A., Xavier Le Roux, Franck Poly, et al. "Plant Species Diversity Affects Soil–Atmosphere Fluxes of Methane and Nitrous Oxide." *Oecologia* 181, no. 3 (2016): 919–30. https://doi.org/10.1007/ s00442-016-3611-8.

The Observatory of Economic Complexity. *OEC Country Profile: Sudan.* Accessed July 29, 2025. https:// oec.world/en/profile/country/sdn.

Oggioni, Silvio Daniele, Raúl Ochoa-Hueso, and Begoña Peco. "Livestock Grazing Abandonment Reduces Soil Microbial Activity and Carbon Storage in a Mediterranean Dehesa." *Applied Soil Ecology* 153 (2020): 103588. https://doi.org/10.1016/j.apsoil.2020.103588.

Oxfam. "Carbon inequality kills: Why curbing the excessive emissions of an elite few can create a sustainable planet for all." Policy and practice briefing paper, Oxford: *Oxfam GB*, (2024). DOI: 10.21201/2024.000039

Oxfam. "Resisting the Rule of the Rich: Defending Freedom Against Billionaire Power." Policy paper, Oxford: *Oxfam GB*, (2026). https://www.oxfam.org.uk/mc/ax2v6v/

Oxfam. "Richest 1% Emit as Much Planet-Heating Pollution as Two-Thirds of Humanity." News release, November 20, 2023. https://www.oxfam.org.uk/mc/qer7km/.

Oxfam. "Survival of the Richest: How We Must Tax the Super-Rich Now to Fight Inequality." Policy paper, Oxford: *Oxfam GB*, (2023). https://oxfamilibrary.openrepository.com/bitstream/ handle/10546/621477/bp-survival-of-the-richest-160123-en.pdf

Pardo, Deborah, Jaume Forcada, Andrew G. Wood, et al. "Additive Effects of Climate and Fisheries Drive Ongoing Declines in Multiple Albatross Species." *Proceedings of the National Academy of Sciences of the United States of America* 114, no. 50 (2017). https://doi.org/10.1073/pnas.1618819114.

Passarelli, Simone, Christopher M. Free, Alon Shepon, Ty Beal, Carolina Batis, and Christopher D. Golden. "Global Estimation of Dietary Micronutrient Inadequacies: A Modelling Analysis." *Lancet Global Health* 12, no. 10 (2024): e1590–e1599. http://doi.org/https://doi.org/10.1016/S2214-109X(24)00276-6.

Pate, John S., and David B. Layzell. "Carbon and Nitrogen Partitioning in the Whole Plant—A Thesis Based on Empirical Modeling." *In Nitrogen and Carbon Metabolism*, 94–134. Springer Netherlands, 1981. https://doi.org/10.1007/978-94-009-8267-3_4.

Pearce, Elena A., Florence Mazier, Signe Normand, et al. "Substantial Light Woodland and Open Vegetation Characterized the Temperate Forest Biome before *Homo sapiens." Science Advances* 9, no. 45 (2023). https://doi.org/10.1126/sciadv.adi9135.

Pearce, Elena A., Florence Mazier, Ralph Fyfe, et al. "Higher Abundance of Disturbance-Favoured Trees and Shrubs in European Temperate Woodlands Prior to the Late-Quaternary Extinction of Megafauna." *Journal of Ecology*,112. (2024): 2813–2827. https://doi.org/10.1111/1365-2745.14422

Pennsylvania Department of Health. "County Profile: Hospitalizations: Maps: Chronic Lower Respiratory Disease," updated February 2025. Accessed August 7, 2025. https://www.pa.gov/content/dam/ copapwp-pagov/en/health/documents/topics/healthstatistics/vitalstatistics/countyhealthprofiles/ documents/current/maps-hospitalizations.html#chronic-lower-respiratory-diseases-1.

Pennsylvania Department of Health. "Health Statistics: County Profile: Bradford: Maps: Hospitalizations." Updated February 2025. Accessed August 7, 2025. https://www.health.pa.gov/topics/HealthStatistics/ VitalStatistics/CountyHealthProfiles/Documents/current/Bradford.aspx, April 20, 2025.

Pennsylvania Department of Health. "Health Statistics: County Profile: York: Maps: Hospitalizations." Updated February 2025. Accessed August 7, 2025. https://www.pa.gov/content/dam/copapwp-pagov/ en/health/documents/topics/healthstatistics/vitalstatistics/countyhealthprofiles/documents/ current/york.html.

Pent, Gabriel J. "Over-Yielding in Temperate Silvopastures: A Meta-Analysis." *Agroforestry Systems* 94, no. 5 (2020): 1741–58. https://doi.org/10.1007/s10457-020-00494-6.

Perri, Angela R., Tatiana R. Feuerborn, Laurent A. F. Frantz, et al. "Dog Domestication and the Dual Dispersal of People and Dogs into the Americas." *Proceedings of the National Academy of Sciences of the United States of America* 118, no. 6 (2021). https://doi.org/10.1073/pnas.2010083118.

Perrotti, Angelina G., Christopher A. Kiahtipes, James M. Russell, et al. "Diverse Responses of Vegetation and Fire after Pleistocene Megaherbivore Extinction Across the Eastern US." *Quaternary Science Reviews* 294 (2022): 107696. https://doi.org/10.1016/j.quascirev.2022.107696.

Pershing, Andrew J., Line B. Christensen, Nicholas R. Record, Graham D. Sherwood, and Peter B. Stetson. "The Impact of Whaling on the Ocean Carbon Cycle: Why Bigger Was Better." *PLOS ONE* 5, no. 8 (2010): e12444. https://doi.org/10.1371/journal.pone.0012444.

Petit, J. R., J. Jouzel, D. Raynaud, et al. "Climate and Atmospheric History of the Past 420,000 Years from the Vostok Ice Core, Antarctica." *Nature 399*, no. 6735 (1999): 429–36. https://doi.org/10.1038/20859.

PEW Charitable Trust, "Ending Harmful Subsidies Could Increase the Amount of Fish in the Ocean," news release, June 7, 2021. PEW Charitable Trust, accessed 1/13/2026. https://www.pew.org/en/research-and-analysis/articles/2021/06/07/ending-harmful-subsidies-could-increase-amount-of-fish-in-the-ocean-research-shows.

Pigati, Jeffrey S., Kathleen B. Springer, Jeffrey S. Honke, et al. "Independent Age Estimates Resolve the Controversy of Ancient Human Footprints at White Sands." *Science 382*, no. 6666 (2023): 73–75. https://doi.org/10.1126/science.adh5007.

Pimentel, David, Alison Marklein, Megan Toth, et al. "Biofuel Impacts on World Food Supply: Use of Fossil Fuel, Land and Water Resources." Energies 1, no. 2 (2008): 41–78. https://doi.org/10.3390/en1010041.

Pimentel, David, and Michael Burgess. "Soil Erosion Threatens Food Production." *Agriculture* 3, no. 3 (2013): 443–63. https://doi.org/10.3390/agriculture3030443.

Plant, Genevieve et al., "Inefficient and Unlit Natural Gas Flares Both Emit Large Quantities of Methane." *Science* 377, no. 6614 (2022): 1566–1571. https://doi.org/10.1126/science.abq0385.

Popkin, Barry M., Camila Corvalan, and Laurence M. Grummer-Strawn. "Dynamics of the Double Burden of Malnutrition and the Changing Nutrition Reality." *The Lancet* 395, no. 10217 (2020): 65–74. https:// doi.org/10.1016/S0140-6736(19)32497-3.

Prates, Luciano, and S. Ivan Perez. "Late Pleistocene South American Megafaunal Extinctions Associated with Rise of Fishtail Points and Human Population." *Nature Communications* 12, no. 1 (2021): 2175. https://doi.org/10.1038/s41467-021-22506-4.

Prates, Luciano, Gustavo G. Politis, and S. Ivan Perez. "Rapid Radiation of Humans in South America after the Last Glacial Maximum: A Radiocarbon-Based Study." *PLOS ONE 15*, no. 7 (2020): e0236023. https://doi.org/10.1371/journal.pone.0236023.

Rees, William E. "The Human Eco-Predicament: Overshoot and the Population Conundrum." *Vienna Yearbook of Population Research.* 21, no. 39 (2021), 21–29. https://doi.org/10.1553/p-eznb-ekgc.

Retallack, Gregory J. *Soil Grown Tall.* Springer International Publishing, 2022. https://doi.org/10.1007/978-3-030-88739-1.

Retallack, Gregory J. "Global Cooling by Grassland Soils of the Geological Past and Near Future." *Annual Review of Earth and Planetary Sciences* 41, no. 1 (2013): 69–86. https://doi.org/10.1146/annurev-earth-050212-124001.

Riemer, Heiko, Stefan Kröpelin, and Andras Zboray. "Climate, Styles and Archaeology: An Integral Approach Towards an Absolute Chronology of the Rock Art in the Libyan Desert (Eastern Sahara)." *Antiquity* 91, no. 355 (2017): 7–23. https://doi.org/10.15184/aqy.2016.252.

Rizzuto, Matteo, Shawn J. Leroux, and Oswald J. Schmitz. "Rewiring the Carbon Cycle: A Theoretical Framework for Animal-Driven Ecosystem Carbon Sequestration." *Journal of Geophysical Research: Biogeosciences* 129, no. 4 (2024). https://doi.org/10.1029/2024JG008026.

Roberts, Callum M., Bethan C. O'Leary, Douglas J. McCauley, et al. "Marine Reserves Can Mitigate and Promote Adaptation to Climate Change." *Proceedings of the National Academy of Sciences of the United States of America* 114, no. 24 (2017): 6167–75. https://doi.org/10.1073/pnas.1701262114.

Roberts, Spencer, Jennifer Jacquet, Patricia Majluf, and Matthew N. Hayek. "Feeding Global Aquaculture." *Science Advances.* 10, no. 42 (2024). http://doi.org/10.1126/sciadv.adn9698.

Rogers, Alan R. "Genetic Evidence for a Pleistocene Population Explosion." *Evolution* 49, no. 4 (1995): 608–15. https://doi.org/10.1111/j.1558-5646.1995.tb02297.x.

Roman, Joe, and James J. McCarthy. "The Whale Pump: Marine Mammals Enhance Primary Productivity in a Coastal Basin." *PLOS ONE* 5, no. 10 (2010): e13255. https://doi.org/10.1371/journal.pone.0013255.

Roman, Joe, James A. Estes, Lyne Morissette, et al. "Whales as Marine Ecosystem Engineers." *Frontiers in Ecology and the Environment* 12, no. 7 (2014): 377–85. https://doi.org/10.1890/130220.

Roosevelt, Theodore. *A Book Lover's Holidays in the Open.* Charles Scribner's Sons, 1916.

Rosenberg, A. A., W. J. Bolster, K. E. Alexander, W. B. Leavenworth, A. B. Cooper, and M. G. McKenzie. "The History of Ocean Resources: Modeling Cod Biomass Using Historical Records." *Frontiers in Ecology and the Environment* 3, no. 2 (2005): 84–90. https://doi.org/10.1890/1540-9295(2005)003[0078:THOO RM]2.0.CO;2.

Ross, Monique, and Ben Cheshire. "Meet the Defiant Grazier Using Feral Donkeys to Regenerate His Land at Kachana Station in an 'Illegal' Experiment." ABC News, November 3, 2024. https:// www.abc.net.au/news/2024-11-04/grazier-using-wild-donkeys-to-regenerate-land-in-legal-battle/104445766?future=true&.

Rouet-Leduc, Julia, Guy Pe'er, Francisco Moreira, et al. "Effects of Large Herbivores on Fire Regimes and Wildfire Mitigation." *Journal of Applied Ecology* 58, no. 12 (2021): 2690–2702. https://doi. org/10.1111/1365-2664.13972.

Royte, Elizabeth. "Livestock Falling Ill in Fracking Regions." NBC News, Nov. 30, 2012. https://www. nbcnews.com/news/investigations/livestock-falling-ill-fracking-regions-flna1c7336300

Ruddiman, William F. "The Anthropogenic Greenhouse Era Began Thousands of Years Ago." *Climatic Change* 61, no. 3 (2003): 261–93. https://doi.org/10.1023/B:CLIM.0000004577.17928.fa.

Ruddiman, William F. *Plows, Plagues, and Petroleum: How Humans Took Control of Climate.* 2nd ed. Princeton University Press, 2005.

Rule, Susan, Barry W. Brook, Simon G. Haberle, et al. "The Aftermath of Megafaunal Extinction: Ecosystem Transformation in Pleistocene Australia." *Science* 335, no. 6075 (2012): 1483–86. https:// doi.org/10.1126/science.1214261.

Saba, Grace K., Adrian B. Burd, John P. Dunne, et al. "Toward a Better Understanding of Fish-Based Contribution to Ocean Carbon Flux." *Limnology and Oceanography* 66, no. 5 (2021): 1639–64. https:// doi.org/10.1002/lno.11709.

Sacks, Adam D., Richard Teague, Fred Provenza, Seth Itzkan, and Jim Laurie. "Reestablishing the Evolutionary Grassland-Grazer Relationship to Restore Atmospheric Carbon Dioxide to Preindustrial Levels." In *Geotherapy,* edited by Thomas J. Goreau, Ronal W. Larson, and Joanna Campe. CRC Press, 2014. https://doi.org/10.1201/b13788-9.

Sanderman, Jonathan, Tomislav Hengl, and Gregory J. Fiske. "Soil Carbon Debt of 12,000 Years of Human Land Use." *Proceedings of the National Academy of Sciences of the United States of America* 114, no. 36 (2017): 9575–80. https://doi.org/10.1073/pnas.1706103114.

Sandom, Christopher, Søren Faurby, Brody Sandel, and Jens-Christian Svenning. "Global Late Quaternary Megafauna Extinctions Linked to Humans, Not Climate Change." *Proceedings of the Royal Society B: Biological Sciences* 281, no. 1787 (2014): 20133254. https://doi.org/10.1098/rspb.2013.3254.

Savory, Allan. "How to Fight Desertification and Reverse Climate Change." TED Talk. March 4, 2013. https://www.ted.com/talks/allan_savory_how_to_fight_desertification_and_reverse_climate_ change.

Savory, Allan, and Jody Butterfield. *Holistic Management: A Commonsense Revolution to Restore Our Environment,* 3rd ed. Island Press, 1999.

Savory, Allan. *The Grazing Revolution: A Radical Plan to Save the Earth,* 2d ed. Savory Institute, 2023.

Savory, Sarah. *Never Lose Sight of Your Elephant,* 1st ed. Published by the author, 2024.

Schipper, Louis A., Paul L. Mudge, Miko U. F. Kirschbaum, et al. "A Review of Soil Carbon Change in New Zealand's Grazed Grasslands." *New Zealand Journal of Agricultural Research* 60, no. 2 (2017): 93–118. https://doi.org/10.1080/00288233.2017.1284134.

Schmitz, Oswald J., Magnus Sylvén, Trisha B. Atwood, et al. "Trophic Rewilding Can Expand Natural Climate Solutions." *Nature Climate Change* 13, no. 4 (2023): 324–33. https://doi.org/10.1038/s41558-023-01631-6.

Schmitz, Oswald J., Peter A. Raymond, James A. Estes, et al. "Animating the Carbon Cycle." *Ecosystems* 17, no. 2 (2014): 344–59. https://doi.org/10.1007/s10021-013-9715-7.

Schulz, John H., David L. Otis, and Stanley A. Temple. "100th Anniversary of the Passenger Pigeon Extinction: Lessons for a Complex and Uncertain Future." *Wildlife Society Bulletin* 38, no. 3 (2014): 445–50. https://doi.org/10.1002/wsb.453.

Shi, Zheng, Steven D. Allison, Yujie He, et al. "The Age Distribution of Global Soil Carbon Inferred from Radiocarbon Measurements." *Nature Geoscience* 13, no. 8 (2020): 555–9. https://doi.org/10.1038/ s41561-020-0596-z.

Shojaeezadeh, Shahab Aldin, Malik Al-Wardy, Mohammad Reza Nikoo, et al. "Soil Erosion in the United States: Present and Future (2020–2050)." *CATENA* 242 (2024): 108074. https://doi.org/10.1016/j. catena.2024.108074.

Siegenthaler, Urs, Thomas F. Stocker, Eric Monnin, et al. "Stable Carbon Cycle–Climate Relationship During the Late Pleistocene." *Science* 310,1313-1317(2005). DOI:10.1126/ science.1120130.

Skerritt, Daniel J., Anna Schuhbauer, Sebastian Villasante, et al. "Mapping the Unjust Global Distribution of Harmful Fisheries Subsidies." *Marine Policy* 152 (2023): 105611. https://doi.org/10.1016/j. marpol.2023.105611.

Slimak, Ludovic, Clément Zanolli, Tom Higham, et al. "Modern Human Incursion into Neanderthal Territories 54,000 Years Ago at Mandrin, France." *Science Advances* 8, no. 6 (2022). https://doi. org/10.1126/sciadv.abj9496.

Smith, Felisa A., Scott M. Elliott, and S. Kathleen Lyons. "Methane Emissions from Extinct Megafauna." *Nature Geoscience* 3, no. 6 (2010): 374–75. https://doi.org/10.1038/ngeo877.

Smith, Vernon L. "Hunting and Gathering Economies." In *The New Palgrave Dictionary of Economics,* 1–7. Palgrave Macmillan UK, 2008. https://doi.org/10.1057/978-1-349-95121-5_1066-2.

Soga, Masashi, and Kevin J. Gaston. "Shifting Baseline Syndrome: Causes, Consequences, and Implications." *Frontiers in Ecology and the Environment* 16, no. 4 (2018): 222–30. https://doi. org/10.1002/fee.1794.

Sokol, Noah W., Sara. E. Kuebbing, Elena Karlsen-Ayala, and Mark A. Bradford. "Evidence for the Primacy of Living Root Inputs, not Root or Shoot Litter, in Forming Soil Organic Carbon." *New Phytologist* 221, no. 1 (2019): 233–46. https://doi.org/10.1111/nph.15361.

Squitieri, Andrea, and Mark Altaweel. "Empires and the Acceleration of Wealth Inequality in the Pre-Islamic Near East: An Archaeological Approach." *Archaeological and Anthropological Sciences* 14, no. 10 (2022): 190. https://doi.org/10.1007/s12520-022-01659-3.

Sremba, Angela L., Brittany Hancock-Hanser, Trevor A. Branch, Rick L. LeDuc, and C. Scott Baker. "Circumpolar Diversity and Geographic Differentiation of mtDNA in the Critically Endangered Antarctic Blue Whale (*Balaenoptera musculus intermedia*)." *PLOS ONE* 7, no. 3 (2012): e32579. https:// doi.org/10.1371/journal.pone.0032579.

Steffen, Will, Wendy Broadgate, Lisa Deutsch, Owen Gaffney, and Cornelia Ludwig. "The Trajectory of the Anthropocene: The Great Acceleration." *The Anthropocene Review* 2, no. 1 (2015): 81–98. https://doi. org/10.1177/2053019614564785.

Sumaila, U. Rashid, Naazia Ebrahim, Anna Schuhbauer, et al. "Updated Estimates and Analysis of Global Fisheries Subsidies." *Marine Policy* 109 (2019): 103695. https://doi.org/10.1016/j.marpol.2019.103695.

Svizzero, Serge, and Clem Tisdell. "The Persistence of Hunting and Gathering Economies." *Social Evolution & History* 14, no. 2 (2015): 3–26.

Taylor, William T., *Hoof Beats: How Horses Shaped Human History.* University of California Press, 2024.

Teague, W. R., S. Apfelbaum, R. Lal, et al. "The Role of Ruminants in Reducing Agriculture's Carbon Footprint in North America." *Journal of Soil and Water Conservation* 71, no. 2 (2016): 156–64. https:// doi.org/10.2489/jswc.71.2.156.

Thaler, Evan A., Jeffrey S. Kwang, Brendon J. Quirk, Caroline L. Quarrier, and Isaac J. Larsen. "Rates of Historical Anthropogenic Soil Erosion in the Midwestern United States." *Earth's Future* 10, no. 3 (2022). https://doi.org/10.1029/2021EF002396.

Thompson, Jessica C., David K. Wright, Sarah J. Ivory, et al. "Early Human Impacts and Ecosystem Reorganization in Southern-Central Africa." *Science Advances* 7, no. 19 (2021). https://doi.org/10.1126/ sciadv.abf9776.

Thomson, Jessame, Sheina Lew-levy, and Duncan N E Stibbard-hawkes. "'Fiercely Egalitarian': Thematic Cross-Cultural Analysis Reveals Regularities in the Maintenance of Egalitarianism Across Four Independent African Hunter-Gatherer Groups." *Cross-cultural Research* Vol. 59, no. 5 (2025): 535– 578. https://doi.org/10.1177/10693971251338210.

UNICEF (United Nations Children's Fund) and WHO (World Health Organization). *Joint Monitoring Program for Water Supply, Sanitation and Hygiene: Progress on Household Drinking Water, Sanitation and Hygiene 2000–2022: Special Focus on Gender.* UNICEF and WHO, 2023.

United Nations. "Climate Action: Causes and Effects of Climate Change." Accessed August 15, 2024. https://www.un.org/en/climatechange/science/causes-effects-climate-change.

United States Environmental Protection Agency (EPA). "Waste Emissions Charge." Accessed January 11, 2026. https://www.epa.gov/inflation-reduction-act/waste-emissions-charge.

United States Department of Agriculture, Economic Research Service. *Food Security Assessment,* GFA-15, 2004, https://ers.usda.gov/sites/default/files/_laserfiche/outlooks/37133/30608_gfa15_002. pdf?v=86566.

United States Department of Agriculture, Economic Research Service. *Food Security Assessment,* GFA-35, 2024. https://ers.usda.gov/sites/default/files/_laserfiche/outlooks/109853/GFA-35.pdf?v=72734

United States Natural Resources Conservation Service, *Soil Survey of Chenango County, New York,* 1985. Accessed January 5, 2026, https://archive.org/details/ChenangoNY1985.

United States Census Bureau. "1950 Census of Agriculture: Volume 2. General Report" (US Government Printing Office, 1952). https://www.census.gov/library/publications/1952/dec/agriculture-vol-02. html.

United States Environmental Protection Agency. "What Are Soil Fumigants?" Accessed May 23, 2025. https://www.epa.gov/soil-fumigants/what-are-soil-fumigants.

United States Department of Agriculture, Economic Research Service. "Legal Status of Hired Crop Farm Workers, Fiscal 1991–2022," infographic, Accessed January 20, 2025. https://www.ers.usda.gov/topics/ farm-economy/farm-labor.

Van der Ent, Rudi J., H. H. G. Savenije, B. Schaefli, and S. C. Steele-Dunne. "Origin and Fate of Atmospheric Moisture over Continents," *Water Resources Research* 46, no. 9 (2010). https://doi. org/10.1029/2010WR009127.

Vera, F. W. M. *Grazing Ecology and Forest History.* CABI, 2000.

Viglizzo, E. F., M. F. Ricard, M. A. Taboada, and G. Vázquez-Amábile. "Reassessing the Role of Grazing Lands in Carbon-Balance Estimations: Meta-Analysis and Review." Science of The Total Environment 661 (2019): 531–42. https://doi.org/10.1016/j.scitotenv.2019.01.130.

Vosburgh, Robert. "The Fall of Promiseland Livestock." *Supermarket News,* July 23, 2011.

Waters, Michael R., Thomas W. Stafford, Brian Kooyman, and L. V. Hills. "Late Pleistocene Horse and Camel Hunting at the Southern Margin of the Ice-Free Corridor: Reassessing the Age of Wally's Beach, Canada." *Proceedings of the National Academy of Sciences of the United States of America* 112, no. 14 (2015): 4263–67. https://doi.org/10.1073/pnas.1420650112.

Waters, Michael R., Thomas W. Stafford, H. Gregory McDonald, et al. "Pre-Clovis Mastodon Hunting 13,800 Years Ago at the Manis Site, Washington." *Science* 334, no. 6054 (2011): 351–53. https://doi.org/10.1126/ science.1207663.

Watters, George M., Jefferson T. Hinke, and Christian S. Reiss. "Long-Term Observations from Antarctica Demonstrate that Mismatched Scales of Fisheries Management and Predator-Prey Interaction Lead to Erroneous Conclusions About Precaution." *Scientific Reports* 10, no. 1 (2020): 2314. https://doi. org/10.1038/s41598-020-59223-9.

Webb, Steve. "Palaeotrophic Reconstruction and Climatic Forcing of Mega-Lake Eyre in Late Quaternary Central Australia: A Review." *Boreas* 39, no. 2 (2010): 312–24. https://doi.org/10.1111/j.1502-3885.2009.00120.x.

Wiercinski, A. "Has the Brain Size Decreased Since the Upper Paleolithic Period?" *Bulletins et Mémoires de La Société d'Anthropologie de Paris* 6, no. 4 (1979): 419–27. https://doi.org/10.3406/bmsap.1979.1979.

Wiessner, Daniel, "N.Y. Top Court Says Towns Can Ban Fracking." Reuters, July 1, 2014. https://www. reuters.com/article/markets/n-y-top-court-says-towns-can-ban-fracking-idUSL2N0P61BT/.

Willerslev, E., Davison, J., Moora, M. et al. "Fifty thousand years of Arctic vegetation and megafaunal diet." *Nature* 506 no. 7486 (2014): 47–51. https://doi.org/10.1038/nature12921

Williams, Curtis. "US Becomes First Country to Export 10 Million Tonnes of LNG in Single Month." Reuters, November 5, 2025. https://www.reuters.com/business/energy/us-becomes-first-country-export-10-million-tonnes-lng-single-month-2025-11-03/.

Wilson, Andrew Ross. "Estimating the $CO_2$ Intensity of the Space Sector." *Nature Astronomy* 6, no. 4 (2022): 417–18. https://doi.org/10.1038/s41550-022-01639-6.

Wolf, Adam, Christopher E. Doughty, and Yadvinder Malhi. "Lateral Diffusion of Nutrients by Mammalian Herbivores in Terrestrial Ecosystems." *PLOS ONE* 8, no. 8 (2013): e71352. https://doi.org/10.1371/ journal.pone.0071352.

Woodall, C. W., B. F. Walters, J. W. Coulston, A. W. D'Amato, G. M. Domke, M. B. Russell, and P. A. Sowers. "Monitoring Network Confirms Land Use Change is a Substantial Component of the Forest Carbon Sink in the Eastern United States." *Scientific Reports* 5, no. 1 (2015): 17028. https://doi.org/10.1038/ srep17028.

Woodson, C. Brock, John R. Schramski, and Samantha B. Joye. "A Unifying Theory for Top-Heavy Ecosystem Structure in the Ocean." *Nature Communications* 9, no. 1 (2018): 23. https://doi.org/10.1038/ s41467-017-02450-y.

World Bank. *Global Gas Flaring Tracker Report,* July 2025, 15. https://thedocs.worldbank.org/en/doc/ bd2432bbb0e514986f382f61b14b2608-0400072025/original/Global-Gas-Flaring-Tracker-Report-July-2025.pdf

World Trade Organization. "WTO Agreement on Fisheries Subsidies Enters Into Force." News release, September 25, 2025, https://www.wto.org/english/news_e/news25_e/fish_15sep25_e.htm

World Food Prize Foundation. "Hunger's Tipping Point: An Urgent Call to Transform Food and Nutrition Security." January 2025. Accessed February 7, 2026. https://www.worldfoodprize.org/index. cfm?nodeID=96854&audienceID=1

World Wildlife Fund. *Living Planet Report 2022: Building a Nature-Positive Society.* Edited by R. E. A. Almond, M. Grooten, D. Juffe Bignoli, and T. Petersen. World Wildlife Fund, 2022.

Worm, Boris, Brendal Davis, Lisa Kettemer, et al. "Global Catches, Exploitation Rates, and Rebuilding Options for Sharks." *Marine Policy 40* (2013): 194–204. https://doi.org/10.1016/j.marpol.2012.12.034.

Wright, David K. "Humans as Agents in the Termination of the African Humid Period." *Frontiers in Earth Science* 5, no. 4 (2017). https://doi.org/10.3389/feart.2017.00004.

Wright, Jonathon S., Rong Fu, John R. Worden, et al. "Rainforest-Initiated Wet Season Onset over the Southern Amazon." *Proceedings of the National Academy of Sciences of the United States of America* 114, no. 32 (2017): 8481–86. https://doi.org/10.1073/pnas.1621516114.

Yang, Yi, David Tilman, George Furey, and Clarence Lehman. "Soil Carbon Sequestration Accelerated by Restoration of Grassland Biodiversity." *Nature Communications* 10, no. 1 (2019) 10:718. https://doi. org/10.1038/s41467-019-08636-w.

Yarrow, David. "Geology into Biology: Carbon, Minerals, and Microbes—Tools to Remineralize Soil, Sequester Carbon, and Restore the Earth." *In Geotherapy*. CRC Press, 2014. http://dx.doi.org/10.1201/ b13788-10.

Zhu, Dan, Philippe Ciais, Jinfeng Chang, et al. "The Large Mean Body Size of Mammalian Herbivores Explains the Productivity Paradox During the Last Glacial Maximum." *Nature Ecology & Evolution* 2, no. 4 (2018): 640–49. https://www.nature.com/articles/ s41559-018-0481-y.

Zimov, S. A., N. S. Zimov, A. N. Tikhonov, and F. S. Chapin III. "Mammoth Steppe: A High-Productivity Phenomenon." *Quaternary Science Reviews* 57 (2012): 26 – 45. https://doi. org/10.1016/j. quascirev.2012.10.005.

Zuckermann, Ghil'ad, and Emmalene Richards. *Mangiri Yarda ("Healthy Country": Barngarla Wellbeing and Nature)*. Revivalistics Press, 2021.

www.ingramcontent.com/pod-product-compliance
Lightning Source LLC
LaVergne TN
LVHW081937200926
848484LV00009B/2451